DOUBLE MOUNTAIN BOOKS

Classic Reissues of the American West

The Big Ranch Country

J. W. Williams

Illustrations by Stephen D. Thorpe

New Introduction by Lawrence Clayton

Texas Tech University Press

The Big Ranch Country is reproduced from a 1971 Nortex Press edition copy and is published as part of the special series Double Mountain Books—Classic Reissues of the American West.

Copyright © 1954 by J. W. Williams

New introduction copyright © 1999 Texas Tech University Press

This book was set in New Baskerville. The paper used in this book meets the minimum requirements of ANSI/NISO Z39.48-1992 (R1997). ∞

Cover design by Tamara Kruciak

Printed in the United States of America

Library of Congress Cataloging-in-Publication Data
Williams, J. W. (Jesse Wallace), 1891-1977.
    Big ranch country / J. W. Williams ; illustrations by Stephen D. Thorpe ; new introduction by Lawrence Clayton.
        p. cm.
    ISBN 0-89672-416-6 (paper : alk. paper)
    1. Texas—History—1846-1950. 2. Texas—History, Local. 3. Texas—Description and travel. 4. Ranches—Texas—History. 5. Ranch life—Texas—History. 6. Frontier and pioneer life— Texas. 7. Cattle trade—Texas—History. I. Title.
    F391.W69 1999
    976.4—dc21   99-13131
                                                            CIP

99 00 01 02 03 04 05 06 07 / 9 8 7 6 5 4 3 2 1

Texas Tech University Press
Box 41037
Lubbock, Texas 79409-1037 USA

1-800-832-4042

ttup@ttu.edu
Http://www.ttup.ttu.edu

THE CHAPTERS

ILLUSTRATIONS

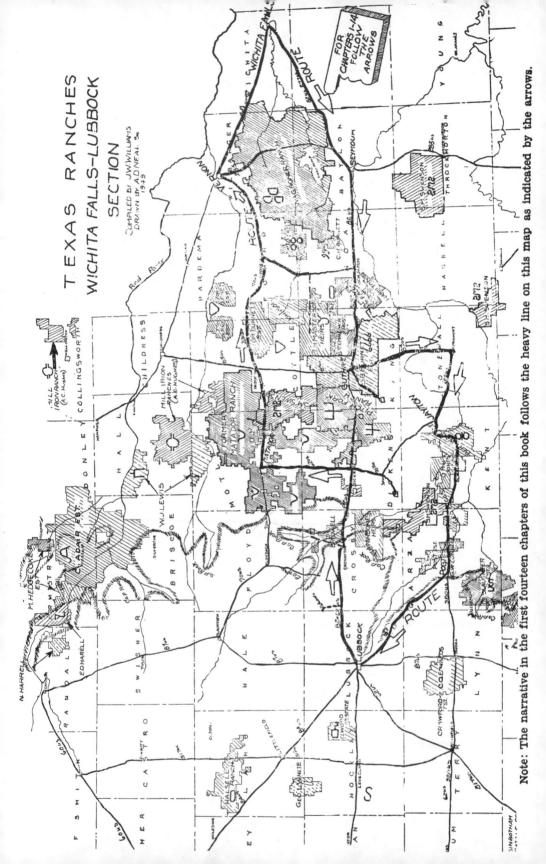

TEXAS RANCHES
WICHITA FALLS–LUBBOCK SECTION

COMPILED BY J.W. WILLIAMS
DRAWN BY ADNEAL 5.
1949

FOR CHAPTERS 1–14 FOLLOW THE ARROWS

Note: The narrative in the first fourteen chapters of this book follows the heavy line on this map as indicated by the arrows.

INTRODUCTION

Texas has a rich ranching history that stretches from the days of earliest European settlement to the present. The history reveals an interesting evolution. The earliest inhabitants, the Native Americans, depended upon meat and skins from wild game, especially bison or "buffalo." Then Spaniards seeking wealth and power brought the first cattle, sheep, and goats to North America as mobile food sources. From these cattle and the horses ridden by the explorers came the two main ingredients for early Texas ranching. Since the Spaniards did not geld male animals, those animals that escaped from the herds or were otherwise turned into the wild became breeding herds. And breed they did on the Coastal Plain that is today the *brasada*, the Brush Country south of San Antonio. Over the decades the animals that escaped Spanish control became feral and wandered at will as their numbers escalated into the hundreds of thousands. To take advantage of the open grasslands and the herds of wild cattle and horses, the Spanish and later the Mexican governments granted huge tracts of property to favored subjects.

Thus, when European immigration came to Texas from the East in the early 1800s, the newcomers found a stable Mexican ranching tradition. Some of the settlers had raised cattle in their Scottish homeland and in the American South, but they found a very different environment in South Texas. The Texicans, as the new settlers were called, learned from and modified Mexican ranching methods to develop what has become Texas ranching. The Mexican *vaquero*, a consummate mounted herder, was the model for the Texas cowboy, a figure that has become mythical around the world. Feral cattle with long horns became the seed stock for the ranchers and produced the millions of animals driven north in trail herds after the Civil War. The animals provided meat for a beef-hungry population back East and seed stock for the Northern ranges. In South Texas, men such as Richard King, Mifflin Kenedy, Francisco Yturria, Thomas O'Connor, and a host of others established huge ranches for cattle, sheep, and goats, mounting their cowboys on mustangs, the descendants of the Spanish horses. The Longhorn and the cowboy became emblems of the era and created one of the most romanticized periods in Texas history.

Ranching in the northern and western parts of Texas developed tentatively at first because of a major hostile threat to the north and west, the nomadic Plains Indians, themselves mounted on descendants of the Spanish horses. The fierce raids of the Indians, mostly Kiowa and Comanche, served as a barrier to settlement. Indeed, early travelers going from Nacogdoches in East Texas to Santa Fe in present-day New Mexico usually chose a route that took them to the coast and then up the Rio Grande to the Mexican settlement to the west, thus avoiding the perils of travel on the desolate Great Plains and the deadly threat of the mounted Indians. Loss of life and livestock was common in early stages of settlement. The situation was eventually changed through military action. A daring raid led by Col. Ranald Mackenzie caught the Indians in Palo Duro Canyon off guard, and in the fall of 1874 destroyed the Indians' winter supplies and their horse herds. This loss and the methodical slaughter of the buffalo by hide hunters ended any hope that the old way of Indian life might be restored.

Thus lay open a vast country, as well suited for grazing herds of cattle as it had been for buffalo, and so remote and arid as to repel any desire for dense settlement at that time. Indeed, even today the area is sparsely populated. The mostly open plains lacked the water and timber resources that had brought Eastern migration to the edge of the woods east of present-day Dallas. Entrepreneurs and future cattle barons seized the opportunity and developed huge ranches across Northwest and West Texas and the Panhandle. Names such as XIT, 6666, Pitchfork, Matador, Waggoner, Adair, Goodnight, Swenson, and a host of others were added to King, Kenedy, Yturria, and others in the southern part of the state. The golden day of Texas ranching had arrived, even as open range grazing and the trail drives to northern areas ended and rail lines spread over the state to provide transportation for moving goods, including cattle, and carrying settlers to remote areas.

Even as the vast expansion of ranching was in progress, there were those who could foresee a time when the pressure of growing settlement and the need for more efficient ranching operations would end the dominance of the huge ranches. There was a need to catalog this grand era, fix it for future generations who would dote on this time and place. J. W. Williams was one of those who answered the call. The result was *The Big Ranch Country*, first published in 1954.

Williams's view was that of a generalist with intense interest in the subject. He was a product of his time, one in which "political correctness" was unknown. Indeed, some contemporary readers may recoil at references to "Negroes" in the role of servants, a common view in Williams's time. Environmentalists will be troubled by the discussion of the wolf hunt, especially that one in Indian Territory when president Theodore Roosevelt and a host of "sportsmen" on horseback hunted the animals down and even caught some of them barehanded. Williams reflected the custom and attitudes of his time. It is easy to fault him now as we fault the lack of conservation practices in the past, especially the slaughter to the brink of extinction of the huge herds of buffalo on the Western plains. Readers must remember that we are all of a time and place, and Williams is unapologetically true to his.

In format, Williams produced a travelogue, much like John Steinbeck's *Travels with Charley* nearly a decade later. Williams wrote of a sparsely populated area beginning to suffer from the big drought of the 1950s that caused people to retreat, especially from West Texas, until the rains came in 1957. He saw ranching undergoing a major transition after World War II, as ranchers mechanized their operations, requiring fewer and fewer cowboys to do the work. Williams's story is anecdotal, its progress often determined by the cooperation—or lack of it—of the car in which the author made the trip. The account is also familiar rather than objective. The reader who seeks dates, market reports, rainfall amounts, and historical facts will find instead lively tales, engrossing anecdotes, personal sketches, and close-up descriptions of roundups, brandings, chuck wagons, campfire meals and stories, and cowboys. Williams's period preceded the current infatuation with the image of the West, embodied in the Marlboro man, that today helps sell items through slick advertising. His study is life as it was lived on the admittedly somewhat barren ranches of Texas. The people Williams met, the sights he saw, and the stories he heard are the fabric of the tale. The informality is perhaps best exemplified in this statement: "After sipping the last drop of my cold drink, I left the drug store and drove out of Post . . . toward Lubbock"(p. 68). Some of us can remember small town drug store soda fountains, but that day is gone, except, like much of the rest of the story in this book, in our nostalgic recollections.

The reader gets lessons in history, geography, and the sources of the names of towns. There are the people—just names in history to us today—who actually ran the ranches. The founders, of course, had already departed the scene. Their influence is powerful, as the photographs of houses and other scenes from the ranches prove. The extensive photograph collection included is a trip in itself. This is a book about a more relaxed day than the present. The cowboys still "lived off the wagon" in the old tradition, not in a pickup truck speeding back and forth from the headquarters to the range talking on a two-way radio or cell phone. The most important information in the book is what I want to call "present history," that is, contemporary information written down at the time for future reference, not a narrative based on archival research. There is plenty of information from the past, but Williams's main contribution in this volume is his description of the world of the 1940s and 1950s, something now slipped away from us.

The day of the big ranches, as Williams describes them, is past. Many of them are gone, the victims of drought, poor markets, poor management, passage from one generation to the next and division among heirs, or lack of interest on the part of the owners or the heirs. It does not always happen that a son or daughter wants to follow the parents in ranching, or has the drive and business acumen to do so. Parents and children may clash, or a ranch may be too small to support the parents and several children. Sometimes the role of running the ranch falls to a grandchild who is the right age to assume control of the ranch when the grandparents are no longer able to run the operation. Some examples illustrate. The Matador Ranch, owned by a Scottish syndicate, liquidated its holdings in the 1950s to recover the financial losses suffered by the Scottish investors in World War II. The ranch was split into several ranches of thousands of acres each. One part still operates under the original name, but is not the vast holding the ranch formerly was. The Reynolds brothers Long X Ranch at Kent in far West Texas has been likewise split up among heirs, but only one large portion is still in the hands of part of the original family of owners. The Swenson Ranch has been split among heirs, and one part of it is now out of family hands. Other large ranches continue to exist and thrive: Spade, Waggoner, 6666, Pitchfork, King, 06. Management is by a corporation on most of these ranches, and they are often run by experts in ranching with degrees, even doctoral degrees, in range management or animal science from such schools as Texas Tech, Texas A&M, Tarleton, Texas Christian University, and Tyler Junior College, all of which have programs in the field.

Ranching will continue in West and North Texas. Much of the region is suited for little other activity. Shallow soil, steep hillsides, deep draws, rocky flats, scant rainfall, absence of large untapped aquifers, and other factors determine that dense human habitation is not in the future for this land. The better soil has been plowed up for farms, and some of the best has been covered with buildings and asphalt. The tracks of horses, cattle, wild game, and a few people are all that much of the big ranch country will know for the foreseeable future. Economics will be a major factor in determining the ultimate destiny of the land. But those tied to the land by their nature or heritage will cling to the traditions of the past and build for the future as ranching—however the future will define it—continues to be part of the life along the Red, the Canadian, the Brazos, and the Pecos rivers and the countless tributaries, draws, creeks, and hillsides on which cattle graze and men on horseback see after the stock. They are all part of a tradition that to most of the world is Texas.

J. W. Williams could not have foreseen the brief time in the early 1980s when sharp increases in the price of oil, long an important supplement to ranching, would bring wealth to many ranchers. Oil allowed for an affluent lifestyle for a time, but it also provided the means to improve pastures by clearing brush, building fences and pens, erecting barns, and expanding water sources. When the price of oil plummeted in the mid-1980s, the shock of vastly reduced revenues devastated many. Only the resourceful survived, as with every crisis—poor markets, droughts, floods, disease outbreaks—faced by ranchers since the beginning. Now income from recreational use of the land, especially from hunting, is replacing oil revenues.

The supposed glamor of the ranching life on remote ranges dissipates when compared to reality. The work was difficult and dangerous. It required skills with wild livestock and the lariat or lasso. The work not infrequently caused crippling or fatal injury. The life was fraught with privation and loneliness, and the remoteness made any human contact more important than it was in the settlements, where ranch workers went infrequently to blow off steam. But to those who led this life, it was a livelihood better than that of the farmer, tied to his acreage as he followed a mule hitched to a crude plow. Those who denigrate the life of the cowboy and rancher should compare it to that lived by others beyond the fringes of civilization. Life was hard everywhere. At least the men on horseback riding the open range, and later the far-reaching fenced pastures, felt a measure of freedom denied many other people of the day. Cowboys still feel that way. Many men left ranching as they grew older, married and took work at some town job in order to raise a family, but they remember their days as cowboys.

As a culture we cling to the time of the big ranches. Jack Pate, a lifetime cowboy, recalled to me in an interview concerning the Long X Ranch in far West Texas, that the large ranches were big enough to "let your feet down and ride." There are men still alive, most admittedly old, who can recall spending nights in canvas-covered soogans, or bedrolls, under the stars for weeks at a time, lulled to sleep by the howls of coyotes and waiting for the "coosie," or cook, to call them to breakfast at the chuck wagon. They still hear in their dreams the sound of the wrangler bringing in a remuda of a hundred and fifty or more saddle horses so that the men could rope their mounts in the early morning half-light for the day's ride. True, no one living can remember the trail drives to Kansas, Montana, and the Dakotas, but plenty can recall the drives to the railroad corrals to load the cattle on cars to carry them to distant markets before the development of the huge cattle vans that today carry cattle from the ranch to market over "trails" paved with asphalt and concrete. Some men recall living in line shacks or distant "camps"—a small house, a barn, a set of pens, and a horse pasture—where perhaps with one or two other men or by themselves they rode out each day to check cattle or water holes as they did what cowboys have always done: take care of business. Many more men can recall life in the bunkhouse as bachelor cowboys in crews of fifteen or more who mounted up at the headquarters and rode miles in the dark to the pastures, rode hard after cattle all day, and then rode home, often in the dark. Their life has been made easier by the eradication of the screw-worm, so that the men do not have to spend long hours prowling pastures looking for infested animals, and by the use of machinery such as working tables and other labor-saving devices.

Unfortunately for the cowboy, each one of these advances and devices allows the ranch to do the work with fewer men. Although the horse remains the emotional center of ranching, the machine—pickup and stock trailer, bulldozer, helicopter, backhoe—does much of the work. In Texas, the horse is in no danger of being replaced by a motorized vehicle because of the brush, cactus, draws, and rough terrain that characterize the landscape on most ranches. Roping the head and heels of a steer from a motorcycle in a pasture seems highly unlikely.

Perhaps the most important development in ranching since the day described by J. W. Williams is the use of the pickup truck and the trailer to haul horses from the headquarters to the pasture for work. Cowboys today spend less time in the saddle than did their predecessors. The negative side of that proposition is that horses get far less work than they previously did and are, therefore, less well trained. But cowboys still work with their horses. In fact, cowboys are misnamed; they are actually horsemen who spend their time taking care of cattle.

In a few years, those who can recall their experiences on the big ranches first hand will have passed from the scene. Anyone who has interviewed them has been enriched by their memories. But cowboys continue to work the ranches of the West, in Texas and elsewhere. We know that as long as cattle roam pastures, ranchers will need cowboys to help take care of the animals. As long as the ranch continues to be an element of life in the American West, and perhaps even if this way of life disappears entirely, some of us will continue to remember this golden day of life in the West. We will revel in the stories of the Big Ranch Country, due in part to the work of J. W. Williams.

I have spent nearly a decade following in Williams's literary footsteps. My interest, like his, is not in the technical details of ranching but in the stories that reveal how these operations came to be and how they continue. I want to know the people who do the work. I first read Williams's work when writing my first volume on ranches in the early 1990s and was intrigued by his view and by the stories. At least subconsciously, I was influenced by his work in doing mine and have continued in that vein.

Lawrence Clayton
Along the Clear Fork of the Brazos River
in what is still ranch country

TO RUTH,

My wife and partner, whose great courage has inspired the best effort that has gone into the make-up of this Volume.

FOREWORD

This book is a narrative of my own journeys through the big ranch country—true down to such little details as the appearance of a quail or a road runner on the highway. A few short side trips actually made at different times have been related as though they were a part of the two principal journeys that make up the book as a whole. Many personal interviews and an extensive searching of records have supplied most of the facts for the story that is told along the road.

Hundreds—in fact thousands—of miles of travel by automobile have been made through the land of big ranches in order to write this book partly as the story of my own personal experiences. Some scholars and critics may object to this informal method of surveying the ranching industry but the person who wishes his reading matter served up story-book fashion surely will not mind.

Maps of many dozens of the greatest ranches are printed on just seven pages of this book. They represent a great part of a year's work. They are not as complete as was my original intention, but they are far more nearly complete than anything of the kind that is already in print.

I hope the reader will realize that the many western stories related in this volume are not in any way re-touched with fiction, except the two or three tales that are related as possible fiction.

My obligations to those who assisted with this work are many Mrs. Ione Parfet did me much service in calling the attention of critics to this book. Professor Foster Harris of the University of Oklahoma has read the entire manuscript. His suggested changes have been adopted. Miss Eva Weber formerly of Midwestern University, rendered similar criticism. Miss Nell Sammons and Miss Louise Kelly of Wichita Falls Senior High School, have likewise constructively criticized the manuscript while it was in preparation. The split infinitives and the dangling participles were largely eliminated by Mrs. Ruth Doty Harris who checked the language mechanics of the entire manuscript. Mrs. Roy Knight did the typing which included much service over and above the normal duties of a typist.

The librarians and their staffs of Wichita Falls High School Library, Kemp Public Library and the Fort Worth Public Library have aided greatly and for the same type of assistance Miss Natalie

Gorin of Wichita Falls and Miss Llerena Friend of Austin are due special thanks.

Mr. and Mrs. Ernest Lee, who have had extensive acquaintance with people and conditions in a part of the big ranch country, have rendered invaluable aid. Mr. Paul J. Pond has contributed a number of excellent ranch house photographs and has spent much time in finishing prints for the photographic section of this volume. Mr. Dick McCarty has likewise aided with ranch photographs and Mr. C. L. Brown has made a number of finished prints. Grateful acknowledgements are due Mr. George Humphreys, Mr. Riley Thacker and Mr. Jim Gibson for old ranch photographs, and to the W. T. Waggoner Estate for photographs of more recent date.

Mrs. A. F. Edwards has assisted with some of the proofreading. Miss Sue Pierce, with supervision by Miss Natalie Gorin, has done most of the index of the book. The contributions by most of the persons interviewed are specially acknowledged in the various footnotes, but for some reason William Graves who aided with a number of the best of the ranch stories has been left out of the notes.

Thus there seems to be no end to the list of persons who have rendered aid.

J. W. Williams

1 — Where the Big Ranches Begin

Two men in a car pulling a horse trailer stopped at the cattle pens a few blocks from down-town Paducah, Texas. Sheriff Calvin Brooks and his deputy drove up beside it and slammed on their brakes.

Joe Meador, who told me this story, said the first two men, whom I shall call Bill and George, were suspected characters, and the officers had been shadowing them for sometime. They had just unloaded a calf from the trailer as the officers came up. The young animal—wild as a buffalo—butted his head into the ground and turned and twisted about over the pen.

"Where did you get that calf?" the sheriff asked.

Bill and George both looked at the ground and hesitated. Finally Bill mumbled:

"We bought him off of Farmer Blank, over west of Cee Vee."

Deputy Wiley Ellis shifted his left foot, eyeing the two men suspiciously. "You sure that's where you got 'im? This calf's a white-face, and Farmer Blank doesn't have anything that would even come close."

Neither Bill nor George had an answer for that, so the officers locked them up for further questioning.

It was late in the day but these two men of the law phoned Mr. M. J. Reilly, ranch manager at the Matador headquarters.

"So it developed sooner than we expected," Mr. Reilly said.

Next morning, Mr. Reilly and the cattle inspector, Ed Russell, came to Paducah. Russell took one look at the calf which bore no brand:

"Look, this calf's nose is still wet," he said, pointing to the slick hairs on the calf's nose, which indicated he had never been weaned.

Both men felt sure the animal was a Matador calf, but here was a chance to prove it. The young maverick was loaded into a trailer and hurried to the north side of the Matador Ranch near Cee Vee. Ed Russell went ahead to see what clues he might find. He heard a cow bawl over near the Fairbury windmill and soon had the calf brought to that point and unloaded.

When the calf's feet hit the ground, he uttered a sound or two and started running in the direction of a grove of mesquites. A cow almost hidden by the mesquite brush bawled and started toward the calf. Branches popped as the two animals raced toward each other. The calf's low bawl was chopped into syllables as he ran. The reunion between an anxious mother-cow and her lonely son was not long delayed.

How the two knew each other may seem like a mystery, but it was enough evidence in the cow country to send a present day cattle thief to the penitentiary.

Joe Meador who lived near the east end of the big Matador ranch explained that the Matador cowboys had begun to be suspicious about the many unidentified tracks around the Fairbury mill. They knew that a number of calves had been getting away from them.

Joe's story hit home too close for comfort. For I, too, had once made the long lonesome drive across that immense Matador pasture near Cee Vee. Thinking of it made me realize that the Matadors were not alone in letting things of value get away from them.

Some of the greatest ranches in the nation, packed with the thrills of the cattle business for more than fifty years, were right under my nose. Rustlers ran their trucks along the very same paved highways over which I drove. And in the not-so- long ago, guns barked in the country near me, to settle the arguments between cowboys and cattle thieves.

Somehow I had not been awake to the things at my very elbows. The West that was all around me was slipping away from me daily—in my scheme of things it was far worse than the loss of calves sustained by the Matadors. People were still numerous who could tell me about the audacity of the rustler Jake Fry or the fine clean neighborliness of Boley Brown. It was only necessary to ride over and ask Sam Graves in order to learn how a man and one of the greatest horses of the west could be pals. A full score of old cowpokes were still at hand who had known this empire of grass almost from its beginning—men who had watched the cattle kings grow and prosper as they made much of the map of the big ranch country as we know it now.

It has been a lot of fun to sit down cross-legged and get the story straight from these old timers and to pass it on to you in this volume. Sometimes I sit at night with my eyes closed and piece their tales together and merge them with a lot that is already written in the records. It's a little like watching a giant movie reel—a reel on which the big ranchers of the West have slipped quietly into view.

A full hundred years ago came the Kings and Kenedys down in the low country; Don Milton Favor out in the highlands; and strangest of all, S. M. Swenson who bought the first great ranch down below the foot of the plains years before he had any thought of ranching! Then came the Civil War to scramble cattle brands and ownerships almost beyond recognition and after that catastrophe the open range had its day of glory for a full twenty years. Then at the mid point of that twenty years, men like Burk Burnett and Dan Waggoner pushed the limits of civilized white men a little farther to the West. Just before the year 1880 cattlemen both small and large, too numerous to mention here, came swarming up toward the plains on the land where the hunters had all but exterminated the buffalo.

On the heels of those native cattle raisers, large syndicates of American and Scotch capital purchased great chunks of this open range and extended the practice of large scale production to the cattle business. The Spur, the Matador, the Prairie and the mammoth XIT are part of the story ahead.

And while we think about these giant ranches coming on the scene—we are reminded that a peculiar race has been in progress. It has been a race without an official starter or a stop-watch or a field judge—a race nobody knew about until it had run most of the way. It was a race for first place among the big ranches. To be sure everybody knows that the X I T was once the greatest of all ranches and that at a later date the King Ranch surged to the front, but don't you be at all surprised when quite a number of pages over in this book we name a brand new champion. The proof came from a lot of sweat and a hundred official land books.

But now the thing in the pages ahead that may be too strange to even square with fiction is the case of a big Eastern breakfast food manufacturer who came a thousand miles away from home and away from his background of successful experience and bought a big ranch. No, he didn't want to raise cows; he intended to cut up the land and sell it out to farmers. Strangely he had hit the key note of the West in the years that followed the turn of the century. The Old West had rounded a new corner. Great stretches of its fertile acres were now ready for the plow.

A new day not without glamour in its own name had come to stay, but this plowing up of these millions of acres of grass might well cause us to wonder about the dwindling supply of beefsteak. Will America soon be forced to half-rations of the meat that accompanies her daily bread? Have patience, and we'll try to find the answer.

In fact I've driven thousands of miles both among the old timers and into the record books to hunt out many of the answers. It's been fun all the way. I ought to buy Joe Meador a good hat.

* * * *

It was on a certain morning in June that I found myself driving southwestwardly out of my home town of Wichita Falls, Texas, headed straight into what was once the Old West. Today's big ranch country began just thirty miles ahead of my front bumper.

It was that country, thirty miles ahead and for some two hundred miles beyond, that had breathed as deeply of the spirit of the Old West as any part of North America. It was that spread of rolling and broken prairies that had beckoned to the plodding farm boy as a great dream country filled with adventure. It was those limitless acres of grass that had lured a sizeable pile of good Scotch money from 5000 miles over the sea to rise and sometimes to fall with the tide of western prosperity. It was that belt of the Great Plains extending at least a thousand miles northward and southward that had lived vigorously when the romance of the Old West was in its golden age.

To be sure, that land has changed from its early state—wire fences, paved highways, railroads, great patches of farm land, new cities, and a dozen cross currents of the modern World have made it anew. But many of the greatest ranches of the West still checkerboard that part of the map and within and near them the old cattle kingdom lives on, side by side with the streamlined age of present day industry, in a strange mixture little known to the average American.

This story, however, does not begin either thirty miles or two hundred miles ahead of the car but in the heart of downtown Wichita Falls. The land that is now in the middle of this western city once served as the roundup ground for a large ranch. In 1877, the area of the city now most sorely afflicted with traffic jams and neon signs was the place where Dan Waggoner branded his cattle[1] and cut them out for market.

The Waggoners themselves, who were first to put the country around the Falls of the Wichita River to the uses of civilized white men, are no small part of my story. Some eighty years ago Dan Waggoner set up his small ranch headquarters just beyond the hill top, three miles east of present Wichita Falls and spread the D 71 herds over many miles of adjoining range. His front gate—figuratively speaking, of course, for there was no fence—was the Van Dorn Crossing on the Little Wichita River fifteen miles southeast of the town of today, and his back gate was Bare Butte, a small mountain ten miles west of town.[2] Except for an occasional trapper or hunter and a few lurking Indians, Waggoner and his cowboys had full possession of this little empire. It was not necessary to own the land nor to do any serious worrying about land taxes or grass leases. The grass was just as free as the water that in favorable seasons lay in pools along the streams, and a cow grazed and grew without much more expense or restraint than a wild buffalo. The one noted difference between a buffalo and one of this herd of cattle was the brand, D71, which Waggoner burned into the hides of his far flung possessions in cow flesh.

It was from Waggoner's historic roundup ground that I began my trek into the great ranches of West Texas.

Ten miles on my way, the aforesaid Bare Butte stuck up its head a mile or two to the right of the road. I think the name

1. Interview with the late W. H. Portwood, who was a Waggoner employee for a number of years.
2. These details about the Waggoner Ranch came from an interview with W. D. (Shinnery) McElroy late of Harrold, Texas. McElroy was a Waggoner cowboy as early as 1879 and for a period of some thirty years afterward. He and his little brother and sister were captured by Indians in Montague County, Texas, at the time that the rather highly publicized Dot Babb was captured. The three McElroy children were ransomed near old Fort Cobb in Indian Territory after about a year of captivity. The price of their freedom was 1800 pounds of powder, lead and flour. In 1875, five or six years after they had rejoined their parents the family moved from Montague County to a point about two miles east of the business district of present day Wichita Falls. Their small herd of cattle shared the grazing area with the much larger Waggoner cow outfit which had occupied this range since 1871.

really ought to be Mount Barry—in fact, it bears that name on an old military map in my collection, and it should be remembered that one Buck Barry, a famous Texas frontiersman, fought Indians over much of the area in the 1860's.

This mound was on the western line of Waggoner's range, perhaps not incorrectly mentioned earlier as his back gate. Dot Babb, one of his cowboys, lived in a dugout near by and kept his D71 cattle from going astray. Such camps, known then as line camps, were placed at convenient points around the range of a cattleman to keep his herds within his selected boundaries. The practice was cow-country custom until wire fences came. Dot Babb, the keeper of this camp, was unique among the early Waggoner cow hands. He, as a small boy, had been carried away into captivity by raiding Indians;[3] and after months of the wild strange life among these nomads of the plains, had been ransomed and brought back to the ways of the white man.

When the Waggoners first came to the Wichita River, Bare Butte became not only their western boundary, it was the last outpost of civilization to the west. Two or three years later, the Ikards moved up the Wichita River Valley[4] and began ranching to the west of Waggoner. Their range covered the immense K. M. A. oil field of today and their cattle bearing the famous V Bar brand covered thousands of acres in addition.

A little log cabin with a rawhide roof, a few hundred yards northeast of what is known as Kadane Corner in the heart of the oil field was the unpretentious headquarters from which Ikard managed this empire of grass and cows that was larger than some of the petty kingdoms of the ancient world.

I drove along the highway past Dot Babb's old line camp into this one time principality of the Ikards — past farms, small ranches, oil fields—through the villages of Holliday and Mankins. Two or three miles to the left, a small creek (Holliday Creek) paralleled my route. Beyond this little stream to the south was the early day range of Harrold Brothers whose Bar X brand once covered most of the grass land in Archer County. By friendly understanding, this small creek served as a boundary between the two great ranches.

By 1884, small cattlemen began to crowd the range, and these

3. From the McElroy interview. See footnote No. 2.
4. This account of the Ikard Ranch is from the unpublished memoirs of W. S. Ikard. The paper is in the hands of his daughter Mrs. Marvin Smith of Wichita Falls, Texas. In 1873, the Ikards began ranching in the area that is now the KMA oil field.

two pioneer ranchers moved their vast herds and occupied a common range in old Greer County,[5] fifty miles to the northwest. Part of the range left vacant by the Ikards and Harrolds was occupied by the T Fork Ranch. According to Mart Banta, who has lived on near-by Beaver Creek since 1877, the owners of T Fork built some two hundred miles of fence to enclose their pasture. Mart can still lead one to some of the old fence, but the T Fork itself has long since gone the way of the buffalo and the longhorn steer.

Quite different has been the story of Dan Waggoner. It is said the T Fork did not bother to own any land: by contrast the Waggoners purchased an empire. They own some 500,000 acres chiefly in Baylor and Wilbarger Counties. As I drove along a few miles past the village of Dundee, I looked out to the right of the road across Lake Kemp and at the heart of this enormous estate that has passed on to the heirs and descendants of this same Dan Waggoner who once counted his cows and branded his calves in the heart of down town Wichita Falls.

To the left of the road and opposite part of the Waggoner lands, the Cowans own a fair sized modern ranch stocked not only with cows but with a small herd of antelope and a few buffalo. Some two or three miles before reaching Seymour, I crossed the famous Western Cattle Trail of other days. For eight years it was the big road to wild and wooly Dodge City, Kansas, and to no end of adventure. Many cowboys and far more than 1,000,000 head of cattle passed that way from 1877 to 1885.[6] One can not even see the old trail now as he drives along the highway but there are many hillside slopes in the back country where it is still visible.

Some ten miles to the north this once busy thoroughfare crossed the middle of Lake Kemp,[7] a modern irrigation reservoir and fishing resort that lies in the heart of the great Waggoner

---

5. The late Lewis Richardson of Jacksboro, employed by the Harrolds in 1884, helped to make the drive to Greer County. An interesting statement gleaned from his personal account was the declaration by Dan Waggoner that a cowman could neither afford to own or lease the land upon which his cattle ranged. However, whether he felt that he was forced to buy land to protect his interests or whether he experienced a change of heart, Waggoner and his descendants became great land owners.

6. The **Fort Griffin Echo**, and the **Albany News** have preserved the statistics of these cattle drives during the years 1879 to 1885. The files of those old newspapers, still preserved at the office of the **Albany News** in Albany, Texas, tell such interesting facts as the ownership of each of these many herds of cattle being driven northward to market the number of cattle per herd, the description, etc.

7. Dr. A. B. Edwards of Henrietta, Texas, who lived near the site of present day Lake Kemp during a part of the time when trail driving was most active, gave the writer this and other bits of information about the route of the Western cattle trail.

ranch of today. Dan Waggoner began to acquire this mammoth range some sixty-five years ago, not a great while before the last of the trail herds raised their long slow moving clouds of dust enroute from south Texas to Kansas.

Waggoner's purchase was typical of the change that took place between 1880 and 1890. By the latter date the open range had almost reached the vanishing point[8] and, with minor exceptions, free grass was no more. As we have seen, syndicates of eastern capital or corporations financed in England or Scotland bought and fenced millions of West Texas acres, and private property in land became a fact instead of a theory. Along with these great syndicates, some of the individual Texas cowmen, such as Waggoner, grew to the proportion of financial giants out of their sheer success as ranchmen.

Dan Waggoner, as if he wished to herald the new age with a different symbol of ownership, changed his brand. The old D71 was discarded and in 1881 the now famous Three D brand[9] was

8. W. C. Holden, **Rollie Burns**, The Southwest Press, Dallas, Texas, 1932, pp.136-37.

9. See Geo. B. Loving, **The Stock Manual**, George B. Loving, Fort Worth, Texas, 1881, p. 52. Note that this directory of cattle brands was published in 1881. The Three D's Brand was not included in this issue. Dan Waggoner still used the D71 Brand on his horses but seems to have discontinued it on his cattle by this time. His cattle brand in this book of 1881 is shown as 'SL", a brand which must have been used for only a short time since there is no mention of it among old time cowboys of this section. The **Texas Almanac and State Industrial Guide, 1945-46** (The Dallas Morning News, Dallas, Texas, 1946, p. 232) states that the Three D's brand began in 1881. Probably the brand was adopted by Waggoner after the information for the **Stock Manual** had been compiled.

placed on the sides of his fast growing herds.   The new brand proved more difficult for cow thieves to change; and even now, as has been the case for more than sixty years, it is far and wide the symbol of a Waggoner cow.

A slight halt was made in Seymour before resuming the journey.

Not far across the Brazos from Seymour were the old Millet Ranch headquarters which had existed as early as 1875,[10] when Seymour was nothing more than a turkey roost . His 20,000 or more head of mixed cattle—longhorns, Durhams, and others—sold in 1880 to Hughes and Simpson[11] who ranched near the site of present Abilene.   The Hashknife brand of the purchasers soon covered the prairies around Seymour and for many miles across the river to the south.   But the Hughes and Simpsons of open range days have long since moved from the Seymour country; and now the heirs of the late W. H. Portwood[12] stand highest among the beef producers of this area.

I headed west on the next leg of my journey, momentarily away from this old ranching section, through excellent farms that at the time needed rain, and on to the narrow little ridge that rises high above both the Brazos and the Wichita Rivers.   The flat and fertile farm lands had given way to cedar covered brakes, and the broad table land had tapered to a mere knife edge above the valleys of the two rivers just wide enough for the paved highway.   It is one of the picturesque spots of West Texas—in fact a miniature painted desert lies on both sides of the road for several miles.   Down in the valley to the right not very many miles, was the upper end of Waggoner's ranch.   I had been riding near or in sight of it for fifty miles!

Benjamin, a rather small county seat town, lay four or five miles ahead; its streets are broad, with parking room to spare, as one should expect in the ranch country.   A cold drink, found at the local drug store seemed to hold the temperature down for awhile.

Resuming the journey westward, the old motor hesitated for just an instant as if to give out a warning of some kind—but then it was only momentary—why worry about it?

10.   The 1875 files of the **Frontier Echo** published at Jacksboro and now on file at the office of the Albany News (Albany, Texas) mention several herds of these Millet cattle en route to the ranch in Baylor County.

11.   J. R. Webb "Henry Herron, Pioneer and Peace Officer During Fort Griffin Days," **West Texas Historical Association Year Book**, 1944, p. 44.

12.   Deceased 1949.

A mile or two from town, the road entered the range of the old C. R. Ranch.[13] It was fenced soon after 1880 with the only strands of barbed wire between Seymour and New Mexico.

Some five miles west of Benjamin, the old car climbed the paved highway to the top of a hill. A half mile south of its summit was the Weatherly Spring, well-known to pioneers. Here early day travelers found the last good water for many miles as they moved west along the old wagon trail that once followed much the same course as the modern pavement.

Quite as interesting as the spring was the pioneer woman who lived near it. Mrs. Weatherly, mother to the early-day cowboys, was to them both confidential adviser and spiritual guide. Once when an itinerant minister was overdue on his engagement to hold a protracted meeting at one of the camping places of the cattle country, she took charge of the meeting,[14] arranging for speaking, scripture reading, singing, and praying until the sky pilot arrived. The late Judge W. M. Moore, for many years a prominent citizen of Wichita Falls, was one of the speakers to whom she turned over the services. He was then one of the cowboys, but with his background of college training and gift of eloquence he measured up well to Mother Weatherly's requirements. The cowboys were already on "the sawdust trail" well in advance of the arrival of the parson.

West of the Weatherly Spring for perhaps fifteen or twenty miles is an area still occasionally referred to as Little Arizona. It acquired the name more than seventy years ago when there was an exodus of cattlemen to the free range of Arizona. When this western movement was in full swing, a number of smaller herdsmen decided to make the much shorter move to this spot of open range fifty miles west of Seymour. Laughingly one of them called it "Little Arizona," and the name given in jest has long outlived the era of free grass. This little patch on the map of west Texas became one of the color spots of open range days. Herds were thrown together on the common grass lands, and at night-fall the cowboys of many brands drank black coffee by the same flickering camp fires. Here the fellowship and even the range songs of the old west flourished to a rare degree. Fifty years before the great western melody, "Home on the Range," became a national favorite, it was familiar music among the cow

13. Frequent interviews with W. M. Moore of Wichita Falls, Texas.
14. The W. M. Moore interviews.

camps of Little Arizona.[15] Passing along the highway even today, one feels glad that this expanse of grass and cedar covered hills which so nurtured the traditions of the early west has not surrendered to the plow. The cowboy is still supreme in Little Arizona, although now instead of a herdsman of ancient longhorns he has become the care taker of cattle of much finer breeds. Until recent years the early-day cowboy who served as pinch hitter for the parson of other days, shared in the ownership of one of the finest ranches of this one time oasis of good fellowship.

To the west and also toward the south, hills and much broken country appeared as a part of the landscape. Rising abruptly from the rest, round shaped Buzzard Peak stood little more than a half dozen miles away. From this dome-shaped mound southward to the Brazos River, the countryside was rough and to the eye appeared wild and free. Far down by the Brazos, although not in plain view of my highway, Kiowa Peak made a local climax to these bad lands. For perhaps three centuries past, wild horses had kicked up their heels and had sometimes lived fat from the many square miles of grass in view of the top of that imposing little mountain.

From some strange turn of events—possibly from some failure to patrol carefully the range—wild horses seemed to linger in the broken country near Kiowa Peak. About twenty-five years ago a few of these pretty creatures still ran at large. One of them in particular attracted attention.

Boog Graves, the late B. T. Graves, one of the top hands for the 6666 Ranch, while riding the rough range west of Kiowa Peak one day, saw this much sought for bit of horseflesh in the mouth of a canyon. To his surprise the horse bore no brand. Actually the animal was a wild horse that had grown up under the protection of the Brazos River Brakes—a beautiful bay stallion!

How badly Boog wanted that wild horse only a cowboy could understand. It was no easy matter to throw a loop over the head of such a wary creature to say nothing of the risk involved. But Boog was not a timid cowboy, and throwing caution to the winds, he put spurs to his horse. He rode straight for the mouth of the canyon and forced the beautiful bay ahead of him. At breakneck speed down the canyon thundered this last of the wild stallions, and not far behind his heels were Boog Graves and his cow pony. Could a mere cow horse overtake this wild phantom of the

15. The **W. M. Moore** interviews.

range? "He must," Boog thought, for the horse was too great a prize to give up.

Here was one of the strangest races ever run in the West. Forty years after wild horses were reputed to have been exterminated from the west Texas plains, a live cowboy on a modern cowhorse and within sight of a great modern ranch was in hot pursuit of a wild stallion.

But the big question was could Boog catch the coveted bay? He urged his pony to its last notch of speed—and then a surprise came! The stallion had run to the dead end of the canyon and could not escape. In a wild impulse the beautiful bay wheeled and tried to run past Boog in the narrow gorge—but Boog was a real cowboy beyond anything the stallion could have anticipated. With a whirl of his lariat this Burnett cowhand shot his loop straight to the head of the prize bay, and for the first time in his wild life the stallion felt a rope around his neck. Boog managed to bring his prize to camp much to the admiration of Burnett cowhands.[16]

In due time the wild horse was brought to understand the ways of a modern ranch and at length he became a part of the remuda of the "Four Sixes." Finally this one-time wild bay became old in the employ of the great ranch and—but that part of the story must wait until a later chapter.

Meanwhile let us return to the highway west of Benjamin in the unique area once known as Little Arizona.

---

16. Informant, Mr. and Mrs. R. Ernest Lee of Wichita Falls, Texas. Mrs. Lee is the daughter of the late B. T. Graves.

2—That Was Hoss Flesh, Neighbor!

The highway began to rise up into the hills of eastern King County, but at this change of terrain the old car began to show signs of fatigue. Or perhaps fatigue is not the right word, for the old motor began to falter at intervals as if some unseen hand grabbed the car and held it back for a split second. The operation was repeated much too often for either the comfort or peace of mind for the driver.

Rugged, cedar covered hills continued on both sides of the road and broad panoramas of landscape and peaceful whiteface cattle spread out for miles back toward the east—but a fellow with a jumping automobile is not at all in the mood to enjoy scenery. A road runner jumped up and scampered across the highway in front of the car. For those not informed, this little fellow is a small western bird that can run like an athlete or fly if he chooses and when occasion demands can fight like a little demon against the worst of reptiles. But even a road runner was poor entertainment for a would-be traveler with mysterious troubles under the hood of his car.

Soon the limping antique and its driver entered the range that Sam Lazurus and associates took over when Little Arizona was in its heydey. Sam and his partners sold their holdings to the St. Louis Land and Cattle Company, a firm of cowmen that continued to do business until the beginning of the present century.

Jim Gibson, now a prosperous farmer and land owner who lives some fifteen miles to the right of the highway remembers well the day when their old "8" brand was burned into the hides of many thousands of King County cattle.¹ The ranch was sold to Burk Burnett in 1900 and with the change of ownership, his famous "6666" brand became one of the great symbols of central west Texas.

The story is told that Burnett won a ranch in a poker game on a hand of four sixes and that he adopted the brand to celebrate his winnings. But stories that circulate even for as many years as has this four six tale often have no factual foundation. The truth about Burnett's brand is that Frank Crowley of Denton County, Texas, sold his little herd of about 100 cows to Burnett² in 1871. The herd bore the 6666 brand which the new owner continued as he grew into one of the cattle kings of the southwest. Burk Burnett moved to Wichita County and prospered in the ranching business along Red River a few miles north of the place where the city of Wichita Falls soon had its beginning. His cattle and 6666 brand spread across Red River into Indian Territory where he made arrangements for perhaps the greater part of a half million acres of grass.³ Shortly after 1900, the authorities at Washington decided suddenly to discontinue the grass leases in Indian Territory, and Burnett was confronted with what might have proved to be a disaster. He went to Washington to see his friend, Theodore Roosevelt who was then President of the United States. Roosevelt arranged an extension of the grass lease that allowed sufficient time for Burnett to work out his problem. Aggressive rancher that he was, Burk Burnett purchased large holdings in Carson County northeast of Amarillo in addition to the ranch already acquired in King County, and the now full grown fortune in 6666 cattle had weathered the storm. It was not poker but a knowledge of cows and an understanding of men that established the present-day kingdom of the Four Sixes— that built it from a small herd of a few dozen cows to the thou-

1. My informants on early King County cattle history were the late Sam Graves Guthrie, Texas; W. M. Moore, Wichita Falls, Texas; Lee Ribble, Vernon, Texas; Jim Gibson, Grow, Texas; and John Gibson, Paducah, Texas.

2. This story of the origin of the 6666 brand has had wide circulation, but it has been denied in recent years by the management of the ranch. See Gus L. Ford, **Texas Cattle Brands**; Clyde C. Cockrell Company, Dallas, Texas, 1936, p. 134. Corroborative testimony as to the purchase of the brand from Frank Crowley was also given to me by W. M. Graham of Matador, Texas, who knew Frank Crowley and knew about his sale of cattle to Burk Burnett at the time the transaction occurred.

3. Informant, the late R. L. (Bob) McFall, Wichita Falls, Texas.

sands of whitefaced aristocrats that today cover a third of a million acres of west Texas land.

In 1907 a railroad was built across the Burnett lands in Wichita County, and a little station a few miles south of Red River was named in honor of Burk Burnett. Some ten years later the small railway stop that bore his name became the center of one of the world's richest oil fields, and in due time the dramatic episodes of the wild scramble for riches that took place at the small town beside the railway station were heralded to the world in the motion picture, "Boom Town."

The interests of Burk Burnett and his descendants are no longer connected with the fortunes of this oil town that once created such a flare of excitement. If the boundaries of the cattle kingdom moved a little west, so did the Burnetts. Today both the ranch in Carson and the one in King County are well centered in some of the best of the cow country.

I drove along—in fact limped along—well toward the heart of the ranch in King County.

The center of this 206,000 acre ranch is the county seat town of Guthrie. It is only a village with a courthouse and a few business buildings. Chiefly it is the site of the headquarters of the Four Sixes with a large commissary and magnificent residence that overshadows all else. The residence is the executive mansion from which the affairs of the ranch are administered.

Bud Arnett, who had been the one and only manager of the old "8" Ranch, continued in his same capacity for the Burnetts. His tenure lasted from the open range days of the old west down to the modern period of ranching. He retired about twenty-five years ago.

At the chief filling station in this town of Guthrie, I slaked my thirst and inquired about an automobile mechanic. I was directed to a garage across the street. Moving slowly, the old car hit on all six as if to mock my misgivings, but the mechanic, wise in the ways of ailing motors lifted up the hood and began to tinker. He examined the "points," gave the car a yank or two backwards and seemingly fumbled around a minute or two with a screw driver. "I think that'll fix you up," he said as he accepted fifty cents for his trouble.

Bewildered at his seeming confidence and efficiency, as I started on the highway south toward Aspermont, I had a few anxious moments. To my surprise the old motor performed as if it had just sipped a fresh drink from the fountain of youth. My car troubles were over, for the old familiar purr was back in the engine again.

A half mile south of Guthrie I turned off the main highway to see one of the most interesting characters of the old West— a man who loved horses and whose very life story is inseparably bound up with the story of a great cowhorse. Sam Graves was not just a common cowhand without concern for his job. He took a pride in his work and a pride in each of the members of his mount.

His love and understanding of horses began in his boyhood and had not abated at the time of this visit even though he was well into his ninth decade of life. Bill Wood who lived on Keechi Creek southwest of Jacksboro, Texas—some sixty miles west of Ft. Worth—raised a smooth short-bodied bay colt, born shortly before 1880 when Sam Graves was only a lad. Sam also lived near Jacksboro and saw the sleek young colt often. He admired the pretty young animal and was not at all displeased to learn that Bill had named the little fellow "Hub." Whether Sam took a special fancy to the name, or whether he on occasion had patted the young bay's nose, or just what little touch of intimacy had attached him to the colt is not known, but somehow he and the little bay soon became fast friends.

Sam wanted to own the horse but money was a scarce article. Finally when Hub was still a young animal, Sam struck a bargain with Bill. He offered to split rails in payment for the horse. Bill set the price at 2000 post-oak rails and Sam accepted. For some time the boy pounded away at the post oaks, but at length he had finished the 2000 rails and became Hub's owner. It was the beginning of one of the finest partnerships between a man and a horse ever known in the cow country.[4]

Hub was still accustomed to the wild free life of a colt and had hardly known the touch of leather. Sam carefully trained him until he lost his sense of fear and became a good saddle pony. Soon the boy and his horse developed that perfect understanding known only to a good horseman and a good horse.

Jack County, where Sam lived, was far out toward the Texas frontier where wages were cheap and money was none too plentiful. A vacancy occurred at a stage stand between Jacksboro and

4. This somewhat extended account of Hub, the great cutting horse, is largely from an interview with the late Sam Graves. The interview of some two hours duration took place in 1947 on the Sam Graves Ranch fifteen miles southwest of Guthrie Texas. His story is partly corroborated by an account of himself and Hub written on the back of a photograph made of the horse and his rider just after their victory at Haskell, Texas, in 1898. The photograph is now in possession of Babe Graves at Guthrie, Texas. A copy of the photograph is reproduced herein.

Weatherford and the boy was offered fifteen dollars per month to fill the position. He saddled Hub, rode over and took the job. A regular wage, however small, was an irresistible temptation to a boy who somehow wished to assert his own independence.

Visitors were rather numerous at the stage stand, for it was in the country where there were frequent movements of cattle. The very herds that were soon to be driven up to King County as the beginning of the 8 Ranch were assembled in Jack County not long after Sam began his duties at the stage stand. One day Babe Arnett, one of the cowboys for this new ranch, stopped at the stage station and offered to trade a large dun colored horse for Hub. Sam was tempted by the appearance of the large, fine looking dun and traded with Babe before he fully realized what he had done. No doubt he regretted having made the trade many times afterward, but it was not the end of his association with Hub.

Bud Arnett, manager of the newly organized ranch, offered Sam a job at $20.00 per month, which was both a promotion and an opportunity to go out west into the new cattle country. The offer was too good to refuse and Sam became a cowboy. The new cow outfit barely had enough horses for use in the drive to King County and was obliged to pull the chuck wagon with steers. Sam was the only cowboy with the outfit with any experience in driving steers and the task of driving the wagon fell to his lot. Bud Arnett the manager, rode Hub. The cattle which had been purchased from various people in and near Jack County were driven west to King County. Some of these cattle had belonged to A. S. Simmons and Son of Weatherford, Texas, and already bore the 8 brand⁵ that was soon to be adopted by the ranch as a whole.

Before the drive to King County began, Sam went by his old home and told his mother goodbye. She gave him some very tempting articles of food to take along and asked where he was going. "I'm going West, just as far as there is any West," was Sam's boyish reply.

Often on the drive the boy, who was soon to begin his strange new life out beyond the 100th Meridian, whacked away at the steers but looked longingly at Hub, his former pony. The cowboys and all their cattle arrived here at the site of this little town of Guthrie on June 19, 1882.

Hub was somewhat overworked on the journey and Bud

5. George B. Loving, **The Stock Manual**, 10. This book of brands was a year before the cattle were purchased for the 8 Ranch in King County.

turned him over to one of the cowboys. All the horse herd (the remuda) belonged to the company, but each of the men in accordance with custom was assigned certain horses for his own individual mount. After two or three months on the new range Sam was able to make a trade and get Hub into his own mount. Then began the real training period of this marvel among cowhorses.

Sam taught Hub to handle cows as carefully as a mother might train her own child. It was at round-up time as a cutting horse that Hub became such an outstanding cow pony. He soon learned how to know exactly which cow Sam intended to cut out of the herd and how to pivot quickly and in the right direction to stay continually between his charge and the herd.

This happy association between horse and rider lasted about three years, but Hub proved himself so skillful that he became widely known. After this training period Bud Arnett again put Hub in his own mount and kept him as his own favorite for some twelve or fifteen years. Reluctantly the boy gave up his cherished cowhorse that had by now grown to a weight of about 1200 pounds and had developed far quicker perception and stronger muscles than the average pony.

Bud grew attached to the horse almost as much as Sam had and there was no end to his pride in the horse's accomplishments.

Once at a roundup Gawk Hensley was having trouble with a steer that persistently refused to be cut out of a herd. Bud was riding Hub at the time and felt that the quick witted bay was equal to the emergency. He bet Gawk $10.00 that Hub could cut the steer out of the herd with the bridle off. Gawk knew that Hub was quite a horse, but feeling that Bud was letting sentiment run away with his judgment agreed to cover the bet.

Then Bud rode into the herd and followed the stubborn animal until sure that Hub understood the assignment. He then leaned forward and slipped the bridle off the horse's nose while in full pursuit. The chase went on for sometime with the steer doing all of his range of quick twists and turns. But for each of the quick movements Hub turned a little faster and constantly kept between the unruly steer and the herd.

By this time every cowboy at the roundup was sitting lazily in his saddle watching the show. The steer, still trying to outmaneuver the horse, darted behind a low clump of cedars—the stubborn animal was still unwilling to be outdone. He darted first to the left and then to the right with Hub on the opposite side of the cedars always quick enough to hold him in check. Then for an instant the steer stood still with Hub glaring at him from the opposite side of the bush.

Bud sat there in the saddle without reins powerless to interfere in this contest between horse and steer. None but a good rider could have stayed on the back of a horse so speedy of movement. Then, quick as a flash Hub jumped through the clump of cedars and bit the steer on the back of the neck. His teeth popped so loudly as they came together that the boys all over the roundup ground could hear the sound. The steer bawled is if somebody were killing him and submissively ran toward the "cut" as he was supposed to do in the first place. The onlooking cowboys let up a yell and Bud Arnett became ten dollars richer. None of those present will forget the incident[6] until his dying day, but Hub continued to amaze the men of the cow country.

By 1890 most of the grazing area of West Texas was fenced into pastures, but the old 8 Ranch had acquired 140,000 acres of the land here in King County and stayed and grew in the cattle business. Bud continued as manager after the day of wire fences and continued to ride Hub until well past 1895. The fine bay cutting horse was at last growing old, and Bud turned him out on grass and water to roam at will during his few remaining years.

Meanwhile Sam Graves went into the cattle business for himself. He chose a site in Little Arizona not far from the 8 Ranch. A few miles to the west of his ranch old Hub was still alive, quietly munching the grass among the cedar brakes of east King County. Bud Arnett in his rush of duties had almost forgotten about the old fellow.

While Hub had grown old, the west had undergone some drastic changes. Wire fences had stopped the common herding of cattle, and old cowhands thus cut off in separate ranches, longed to swap yarns with their old buddies again. To supply this renewal of fellowship for the cattle country, a cowboy reunion with horse races and other attractions was announced to be held in 1896. According to contemporary estimates 10,000 people packed the little town of Seymour, Texas, where it was held. It was repeated in August of 1897 with equal success and then the big event was moved to Haskell for the following year[7] and a most interesting feature was added. As a special attraction a $150.00 prize was posted for the best cutting horse in a contest to be held at the reunion.

6. Informant, Lee Ribble, Vernon, Texas.
7. **Haskell Free Press**, August 20, 1897.

Sam heard about the contest and wondered about old Hub. The longer he thought about the big show at Haskell the more his mind lingered on the bygone days with his old pal. He just had to find out about his old horse.

One day Sam rode over to the 8 Ranch to see Bud and to ask about old Hub. Bud said, "Sam, he's over in the cedar brakes on the east side, but he's too old and poor to enter a contest." Sam was anxious to see his old horse again, regardless of outcome, and insisted until Bud consented for him to take the horse and try him.

Then Sam found Hub—a pitiful remnant of the fine horse of other days. His ribs were plainly visible—Sam could have counted them, and the poor old horse's hip bones stuck out almost like hat racks. The one time fine, sleek cutting horse of the old 8 Ranch was twenty-two years old and apparently was about ready to die. But as Sam sadly observed the old horse, he remembered him as a frisky young two-year-old colt again. He remembered the old fellow as the envy of the open range and decided to give him a chance to revive his former glory.

Old Hub was taken to Sam's ranch and fed on soaked oats and prairie hay. He was turned out in a small pasture part of the time for grass and exercise, but he was handled with the fine discerning care of his great trainer. After a few days Hub's ribs began to disappear and his flesh rounded smoothly over his hip bones again. He was beginning to look like the beautiful bay that Sam had known for twenty years. After some ten days of intensive feeding and care, Sam tried him on a little "cow work." The old horse was beginning to "come alive." He could cut a yearling out of a herd with the old-time fire and speed again, and Sam knew in his heart that his old pal was to be one of the top contenders in the contest at Haskell. Ten more days of tender care and good food and it was time to start to the reunion. Sam made the trip in a hack and led Hub behind the vehicle. Made in slow easy stages the journey required two days.

The reunion took place July 27-29, 1898.[8] Every effort had been made to publicize the big event. Both the Kansas City Star and the Dallas News told of the cowboy convention through their columns weeks in advance.[9] Special railroad rates were provided

---

8. Haskell Free Press, July 30, 1898.

9. Some of the liberal publicity of these large newspapers may be found reprinted in the **Haskell Free Press** of June 25 and July 16, 1898. All issues of the **Free Press** from June 25 to July 30 were super-charged with publicity about the cowboy reunion.

from everywhere in Texas. Spectators came from all directions, even though Haskell was some fifty miles from the nearest railhead.

Thousands of people came in buggies, wagons, hacks and on horseback until the little frontier town of Haskell was packed to its very limit. Some estimates placed the crowd as high as 15,000 people. Cowboys were on hand from many miles around and the old days when West Texas was one big 50,000,000-acre pasture without a wire fence to mar it were discussed many times.

Rodeo and racing events by day and band concerts and a popular frontier circus by night entertained the throng of Texas pioneers almost around the clock for two days.[10] And then the third morning arrived with the cutting horse contest as the feature event of the entire reunion. The 400-foot grandstand, that was set down upon the flat prairie near the little town, was filled at an early hour, and wagons and buggies filled with people were soon parked most of the way around the mile-long race track. Everybody was on hand to see the cutting horses perform.

The rules of the contest permitted each entry to spend five minutes during which time he was to cut as many cows out of the herd as possible. Each cow was to be driven to a certain predetermined point where two mounted men were ready to receive the animal. Other things being equal, the cowboy who could deliver the most cows one at a time to the two mounted men was to be declared winner and was to receive the $150 prize money. The judges were also to give due consideration to the skill of both horse and rider.

The time arrived for the contest and eleven cowboys were within the enclosure. Boley Brown from Kent County, one of the eleven, was there on his big six-year old sorrel horse. Boley was a great sport and everybody was betting on his sorrel to win. The gate was finally closed, but in a moment a cowboy from Childress rode up on a fine-looking horse and asked to be admitted as one of the contestants. The gate keeper refused to let him in on the ground that the entry time had closed. A friendly argument followed and Sam rode over to the gate.

"What's the matter?" Sam called out to the would-be contestant.

"I've got the best horse in the country, and I've just lost

---

10. The story of the Cowboy Reunion and most of the details of the contests are from the Haskell Free Press of July 30, 1898. The facts of the cutting horse contest are supplemented by my interview with Sam Graves and by accounts from pioneers who saw the contests.

$150.00 if they don't let me in," said the mounted man from Childress.

"If he's that good, I want to see him work, but remember I've got the best horse in the whole United World," was Sam's friendly banter.

Boley Brown rode over to the gate and insisted that the cowboy from Childress be admitted, and the gate was thrown open.

Then the signal was given and the first entry was called. It was Boley Brown on his fine sorrel. There was a momentary hush and then a yell went up as Boley rushed into the herd. The Little bunch of cattle ("the roundup") from which each cowboy was to make his cuts consisted of a number of spayed heifers fresh from the range and as wild as bucks.

Boley and his fine sorrel did effective work among these unwilling animals. He cut out nine of the stubborn heifers before his time expired, but one of the nine eluded him and never was delivered to the two horsemen. This was a mark against Boley's record that the judges would have to reckon with before announcing the decision.

The next contestant was the man from Childress who had entered the gate at the last moment. He, like Boley, dashed into the herd and cut out a heifer, but the frightened animal got away from him and could not be brought to the delivery point. The exact number of "cuts" made by this second contestant has been forgotten but his total was not high enough to approach the winning mark.

Sam and Hub came third in the order of contestants. When the signal was given, he urged Hub into a good pace and rode straight into the middle of the herd. Sam's legs were long and Hub was low and stocky built. Observers say Sam's feet almost seemed to drag the ground as the short-bodied bay did his repertoire of antics. The first objective was a much-frightened brindle heifer. Hub soon had her out of the herd and then she made a quick side movement, but Hub was ahead of her. Again she tried to turn aside, but Hub was again between her and the herd. The heifer tried again with similar results and in a moment she was delivered safely to the two men on horseback. Next Sam and old Hub turned back for a second heifer and soon delivered her; then a third was put safely away, then a fourth, a fifth and a sixth. Many cowboys who were expert horsemen, with their eyes admiringly following every movement of the old horse, stood up in their seats entranced by his unbelievable skill. Sam's brilliant horse was too old to match the great bursts of speed of the younger mounts of his competitors, but with split-second man-euvers, always ahead of his charge, his performance was compara-ble to the broken field running of a great ball carrier in modern football. He even blocked an obstinate animal with a front foot when necessary.

A yell came from the grandstand and the mile-long circle of vehicles as Sam and Hub brought the seventh heifer to the deliv-ery point; hats went into the air when the eighth animal was put away just before the time expired. Old Hub had come back almost from his grave to thrill this great throng of horselovers almost to the point of madness.

Another contestant, because of the sheer leg power of his young mount, tied this record of eight, but as to skill of horse and rider did not even approach the work of Sam and Hub. This was undoubtedly the decision of the mass of cheering spectators; it was also the decision of the judges who awarded the $150 prize money to Sam Graves.

After the contest was over, Harry Daugherty, president of the Cowboy Reunion, persuaded Sam to cut out a heifer without the use of the bridle. The old horse performed for Sam almost as well as on the day that he had won the ten-dollar bet for Bud Arnett. This time, however, he made two trips into the herd of ner-vous cattle and both times delivered his charge promptly. The crowd roared its approval as Sam and Hub finished the difficult assignment. The twenty-two-year-old horse that had been turned out into the cedar brakes to die had had his final day of triumph.

After the reunion, Hub was taken back to the 8 Ranch, and Sam turned over half of the $150 prize money to Bud. Now the grand old cowhorse need no longer waste away to skin and bones, for Bud's share of the winnings was set aside to supply oats ex-clusively for Old Hub.

But we must leave the story of Hub and come back to my own journey along the road a half mile south of Guthrie. Turning in a southwesterly direction off the highway, I was soon driving among white-face cattle that were grazing on part of the immense pasture lands of the 6666 Ranch—which is the successor in own-ership to the same ranch to which old Hub had belonged when Sam rode him to victory at Haskell. After seeing some ten miles of 6666 grass and cattle, I crossed a cattle guard into smaller tracts of land. It was five miles farther to the ranch home of Sam Graves, the same old time cowboy who had trained Hub.

It was forty-nine years after the cowboy reunion at Haskell. Sam was now eighty-four years old but he was still a lover of fine horses. When I arrived at his home, Mrs. Graves had me wait in the living room—Sam was out in the pasture riding old Button, hoping to find a yearling that needed his attention. After an hour or two I caught a glimpse of Sam on his sleek black cowhorse. He came back to the house at an easy gait, old in years but still at heart a cowboy with the spirit of the old West in his very bones.

I could not resist the opportunity to take the picture of Sam and Button—one does not often point a camera at such a great trainer of cowhorses.

The life story of Sam Graves was full of much that was alive and interesting on the old frontier. I noted as much as possible during an interview that lasted nearly two hours but, because of limitations of time, omitted many colorful details. It is to be hoped that someone will yet devote the space of a long article or a book to the life of this great cowboy who not only trained fine horses but did far more than one man's share toward establishing law and order in the West.[11]

It was soon time for lunch, and here I enjoyed a fine meal and a type of hospitality that is fast passing even from the great Plains. It would have been little short of an insult to these westerners for a guest to have driven away at meal time without joining the family at the dining table. It was an experience that I shall long remember and cherish. My drive back to Guthrie seemed easy because of this fine association.

11. Sam Graves died at his ranch in 1949 since this interview.

3—In The Kingdom of Burk Burnett

After leaving the Graves' home, I intended next to visit George Humphreys, manager of the immense King County unit of the 6666 Ranch. George is widely known for his very human attitude toward his men and for his gracious manner as a host.

As I drove back into the town of Guthrie, I cast a glance backward into the story of that unique little village and the cow country around it. It is possible that some of General Mackenzie's scouts may have visited this flat spot of prairie as early as eighty years ago; but if so, the visits were only incidental. For perhaps a thousand years before Mackenzie's day black herds of buffalo had rolled over this little spot of prairie.

But the story does not become a story with real names until about 1875. In that year John R. Cook, Charlie Hart, Limpy Jim Smith, Charlie Rath and a whole host of other buffalo hunters and hide buyers came down from the north[1] across King County. The place where Guthrie is located was not far off their path. That the hunt became intense in the nearby country is attested by the great piles of buffalo bones left behind. On Willow Creek, northwest of this present day village was once a group of dugouts

1. John R. Cook, **The Border and the Buffalo**, Crane & Company, Topeka, Kansas, 1907, pp. 110-150, 329-338, 348-350.

of the hunters where night life for a brief season almost approached that known to the wild days of frontier Fort Griffin.[2] Other hunters were camped in the valley of the north Wichita, north and northeast of town, and—Charlie Bird and his young wife[3] hunted so successfully eastward of the present village that the little stream where they camped is still known as Bird Creek. But the whole story of the buffalo hunt lies in the indistinct past both of Guthrie and King County almost as intangible as Greek mythology. The buffalo hunters left, nearly to a man, when the hunt was over, but far more of the cattlemen who came after them stayed in this new country and their story is much better known. Some little has already been told about the old 8 and 6666 ranches and the more extended account of ranching in central West Texas is forthcoming in succeeding chapters.

Here at Guthrie is one of the focal points of the old West. Here in the little bit of flat country where the present-day tiny town has grown up, was the old roundup grounds known as far back as men herded cattle in this part of Texas. It became a great meeting place for cowboys for some miles around—it was one point where the fellowship of the frontier often flowed over the coffee cups. To the old-time cowboys it became a real place long before there was any hint of a town. Then in 1891, when the movement to organize King County was under way, the cowboys, perhaps out of pure sentiment, were able to do something about their regard for the little spot of prairie. Officials and stockholders high in the ownership of the old 8 Ranch were anxious to have the new county seat located a little east of the center of the county on Ash Creek; but when election day came, the cowboys held most of the votes and the courthouse was located on this little flat now known as Guthrie.[4]

This organization of King County was a rather strange procedure. In an area so thinly populated it was difficult to find enough actual residents of the county to sign the required number of names to the petition that requested a county organization. Every traveler who went west across King County—whether by covered wagon or on horseback—was shown the petition and

2. Informants, Sam Graves, John Gibson, Jim Gibson, and W. M. Moore. This was the site of the McKamy Store, located on land that now belongs to the Pitchfork Land and Cattle Company.

3. Informant, the late Mrs. Dumont of Paducah, Texas. Mrs. Dumont was the wife of Charlie Bird and one of the few women who actually witnessed a buffalo hunt.

4. This story about Guthrie and the organization of King County was supplied by W. M. Moore, Wichita Falls, Texas.

asked to sign it. But still the list of names grew slowly, and frantically those in charge looked about for ways to complete the petition.

There were in those days two principal ranches that maintained headquarters within the county, the old 8 Ranch already mentioned and the Q. B. Ranch. The headquarters of this latter ranch was five miles southwest of the "8" headquarters. There was friendly rivalry between the two cow outfits, and nobody knew when a prank was likely to be played by one outfit on the other. It was something like an exchange of punts at a football game.

There was even rivalry in their efforts to get the petition signed. The Q. B. boys had a dog named Sam that constantly stayed at headquarters even when all the hired hands were away. His name was added to this petition to organize King County—he was signed as "Sam Householder."

The "8" cowboys did not wish to be outdone and added a signature, John B._____, which was the name of an old burro of the male variety that grazed the range with the "8" cattle. Thus padded even with the names of animals, the petition was completed and in due time King County was organized.

With very little farming around it, Guthrie, the new county seat, never grew past the size of a village; but it has remained with its unique ranch atmosphere all through the years.

The 8 ranch purchased the holdings of the Q. B.'s in 1892 and kept the headquarters of the combined property ten miles east of Guthrie. However, after the ranch sold in 1900, Burk Burnett, the new owner, began to put more emphasis on the town of Guthrie.

Outlying camps were placed at points on the ranch—the South Camp near the southwest corner of the ranch, the Taylor Camp in the southeast corner and the North Camp over in the northwest corner. Finally when Burk Burnett decided to move the headquarters, the old place that had long served as the administrative center of the 8 Ranch became just a camp.

Then one of the big events in the story of the 6666 Ranch took place. Colorful Westerner that he always was, Burk Burnett decided to build the finest ranch home in West Texas.[5] It was 1917 when the immense rock ranch house was completed on the hill west of Guthrie. It cost $100,000 and probably could not be dup-

5. Informant, George Humphreys, Manager of the King County division of the Burnett Estate.

licated today for less than a quarter of a million. Undoubtedly
Burk Burnett had finished the finest ranch house known on the
Texas plains. The beautiful building was not constructed after
the usual architectural plans of ranch houses. It overlooks the
little town of Guthrie today almost with the dignity and a little
of the appearance of a castle on the Rhine.

In company with a friend I drove on through Guthrie and
up to the big rock house on the hill, through the open gate in
the iron fence that surrounded this unique ranch home. Parking
the car almost at the front door, we knocked; but the sounds of
our persistent efforts died in the front part of the big house far
short of any human ear within. At length we went around to
the back of the house and were surprised to find Mr. and Mrs.
George Humphreys doing the dinner dishes. The negroes who usually looked after that chore were away for the day. George wore
a kitchen apron gracefully; and although he was both sheriff of
King County and local manager of one of the World's great
ranches, those weighty honors for the moment seemed to trouble
his brow not a bit. He was host in the open free manner of the
cow country—and Mrs. Humphreys was hostess in similar style.

No, they refused to let us help with the dishes, and a cold
drink was served and chairs pushed forward for us to be seated
right there in the great ranch kitchen while the dishwashing
proceeded. A giant refrigerating unit and an immense drain
board and sink impressed one with lavish manner in which the
builder of this great western home had provided facilities for
those who were to look after his interests.

Soon the dishes were put neatly away, and we proceeded toward the living room—through a spacious rectangular dining
room with a long dining table that had room to accommodate
twenty or thirty persons. The table seemed almost lost—standing out in the center of the big room.

Only a few more steps took us to the living room that seemed
to reflect the pride of Burk Burnett, the former cattle king. A
fireplace that was certainly anything but midget in size attracted
our attention first. Here, well up on the front of this center of
wintertime cheerfulness, was a large white stone with the 6666
brand chiseled neatly into its face. The sixes were a little different from the ordinary sixes that are found on the printed page.
The small loop of each figure is left slightly unfinished just as the
cattle king himself had wished it and just as the brand had stayed

throughout the eightly-three years that it had represented the ever-growing Burnett fortune.

To the left of the fireplace was the picture of a very interesting woman. It was the grand-daughter of Burk Burnett, the daughter of Tom Burnett, and now the chief living heir to these vast Burnett properties. Perhaps she is best known as "Anne" in the cattle world; but to the Cowboys she is "Miss Anne," a title that carries with it something of a mixture of respect, affection and reverence. She is Mrs. R. F. Windfohr of Ft. Worth, Texas— one of the wealthiest cattle women in the world.

There were plenty of large well upholstered, comfortable chairs in the living room; but we did not sit down in them—it was a little too warm for that. It was cooler on the screened front porch that was itself larger than most living rooms. We passed through an arched doorway in the stone wall—probably large enough that one could have driven a car through it. This great opening was flanked on either side by arched windows of similar proportions—windows that came to the floor and that seemed to make one large entertainment area of both living room and front porch.

We sat by a wicker table and examined the many photographs from the old ranch albums that George brought forth. Some of them are reproduced elsewhere in this book—for all of them were offered freely for publication.

Our stay was rather long and punctuated with much talk about cattle, old cowhands—and the subject that always ranks first in the cattle country—"horses." Some of the things about horses that are told on other pages of this volume were learned from George Humphreys here as we sat by the wicker table.

I glanced back at 'Miss Anne's" picture in the living room— she comes to the ranch in person quite often. Not long after my visit she came and brought her whole retinue of servants. It was a great occasion, the wedding of Miss Celia Ann, the daughter of Mr. and Mrs. George Humphreys to Fred Martin of Spur, Texas. Celia Ann always had been a favorite of the Fort Worth cattle heiress and now the time had come to celebrate that fondness for this talented young belle of the west. Most likely the great house built by Burk Burnett had never seen an occasion of greater color or gaity during the thirty years that it had stood on the hill top west of Guthrie. It was truly a wedding of the old West with its silks and satins and cowboy boots, and refreshments at the reception served in pomp and splendor such as might have done credit to some of the old world kingdoms. One thing seems to have been purposely omitted. Newspaper reporters and photographers seem not to have been notified, otherwise this gala occasion would have been spread across eight columns in many Sunday supplements.

The fine old western atmosphere of this ranch house, whether on such special occasions or in the run of everyday, is in a measure matched by the range practices of the Four Sixes. Cowboys who have ranched from Montana to Texas seem to doubt that there is any other spot where the old West of fable and story is preserved quite so well as here within the little kingdom set up by Burk Burnett.

On leaving this cattle mansion and the fine company of the Humphreys, we threaded the curved driveway that wound through the yard and left by an iron gate different from the one through which we had entered. We had intended to visit "The Wagon" while at Guthrie, but it was too late in the day for that now. The Wagon was the term applied to the supply or chuck wagon that served as the ever-moving headquarters for the cowboys who worked the range. The wagon was the place where the cowboys slept at night—the same old improvised headquarters on wheels that had held forth seventy-five years ago when cattlemen first came to the Texas plains. But it was growing late now —we could visit the wagon tomorrow.

Accordingly we drove back nearly seventy miles to Seymour to spend the night. Next day we turned west again toward Guthrie stopping for a while to visit Mr. and Mrs. Brannin—affectionately known as Uncle Dick and Aunt Alice—who lived on the 6666 Ranch until a few months ago. We shall return to their unusual story a few pages farther on.

Later in the morning when we reached Guthrie, we found that the wagon had just moved to McCarrolls Pens eight or nine miles south of town. Driving southward, we soon found the dim prints of wheels along the pavement. On closer examination we discovered two kinds of wheel prints that had been made just ahead of us—two large ones and two small ones. Since we were not far behind, it would not require long to find out exactly what we were trailing.

Soon we crossed Croton Creek, and within a mile beyond were the McCarroll Pens. The Wagon, the wheel prints of which we had followed, had already moved out into the pasture a few hundred yards to the right of the road where the cook, who drives the wagon, had begun to make camp. This place would be the temporary home of the 6666 cowboys for a few days while the round-up was completed in the adjoining part of the ranch.

As we approached the gate through which the wagon had just passed, the highway in front of us was momentarily blocked. The cowboys were driving a small herd of whitefaced cattle across the road toward the branding pens on our right. Shortly, these cattle were penned and the ten cowhands who had just done the penning wheeled back in our direction. While they were about it, we had driven inside the pasture and made ready with our cameras. We asked the boys to hold their position for a moment while we snapped a picture or two. They accommodatingly held the line also while we noted their names in order. These boys were the greater part of the man power necessary to carry on the daily routine on this 200,000 acre ranch. Jack Spencer, a 6666 cowboy of long standing and now the wagon boss, was one of the ten. For those who might misunderstand—the wagon boss is the man in charge of the field crew which has the chuck wagon as its headquarters. He in no way looks after the wagon itself—that comes within the cook's duties.

The other boys besides Spencer were Riley Thacker, Toar Piper, Shorty Freeman, Shirley McClaren, Jake Luttrell, Ted Wells, John Dotson, Bill Mason, Latham Withers and Wayne Piper. Not in the picture but elsewhere on duty at the time was J. J. Gibson, the son of the Jim Gibson mentioned earlier who had served many years as a cowboy on the old 8 Ranch before it became the property of Burk Burnett. Young Gibson who is now with the ranch is a unique figure among cowboys. He is a college graduate with his own plans for a career as a rancher.

Other members of the personnel were two whose names we missed, the horse wrangler and the cook.

And now by an odd coincidence while calling the roll of employees, we were brought face to face with the solution of the mystery about the second pair of wagon tracks which we had just trailed along the pavement. The second and smaller wagon always associated with the chuck wagon is a vehicle known as the "hoodlum wagon." This seeming oddity is devoted chiefly to the gathering of wood and water, and the boy who drives it is called the "hoodlum." In this case the name of the hoodlum escapes us, and for a very good reason his photograph is also missing. With all the steadfastness of a big time capitalist who dreads adverse publicity, this 6666 hoodlum refused to appear in front of a camera

—but nevertheless he constituted the fourteenth member of Jack
Spencer's hired hands and completes the roll.

Lunch was nearly ready and the boys all rode in toward the
wagon, leaving the cattle penned until a greater number could
be brought together. It would hardly pay to heat the branding
irons for so few. We met the boys whom we had not previously
known and engaged in informal talk and snapped a few photographs while the cook finished with the meal. Also some of the
boys stretched up the "fly" which was made of heavy ducking
and served as a tent top for shelter and shade both to the sides
and behind the wagon. The bed rolls, covered with water proof
ducking, had been pitched out of the wagon on the ground, but
this newly stretched fly would protect bed rolls and every thing
else beneath it in case of rain, and although it had been dry for
a long time, there are no sure rules about rain.

The meal consisted of roast beef, sour dough biscuits, red
beans, stewed raisins, canned tomatoes and strong black coffee.
If the cook yelled the traditional "Come and get it," his voice
failed to carry the fifty yards or more out to the car where Jack
Spencer and I were engaged in conversation for the moment. Jack
had come out from the wagon with me to see some fifty-year-old
photographs of old cowboys that once worked for the 8 Ranch.
Some of the boys at the wagon yelled at us to come on and look
after our interests, and their friendly advice soon got results.

I picked up a tin plate and filled it after the manner of the
cowboys—then moved aside a little and squatted ranch fashion to
eat as the others were doing. It did not require very long for me
to discover that my legs were not as adaptable to this kind of
squatting as were the toughened legs of the cowboys. Nevertheless by shifting position several times, I managed to get along
without serious discomfort, but probably an experienced cowhand
could have picked me out as a tenderfoot by the time he had come
within a half mile of the wagon. However, the 6666 cowboys were
a very friendly, tolerant lot and seemed not to notice my unortho-
dox style of posture.

After lunch as the wrangler began to round-up the remuda,
I asked Jack Spencer if there was any unusual horse among this
herd—one that had done anything of note. As a result of the
question I soon learned something about both horses and cowboys. Jack insisted that there was nothing unusual about any
of the hundred or more horses that were being assembled in the
rope corral in front of us. But young Gibson, who had overheard

our conversation, told Ernest Lee, my traveling companion, about the achievements of "Sailor," a black horse that was on the far side of the remuda. Sailor had won second place in a "cutting horse" contest at Pampa, Texas, in 1945 and a fourth place in 1946, and he had won first in a similar contest at Paducah in 1945. It was none other than Jack Spencer who had ridden this fine horse to victory in all those contests, and his failure to tell about this excellent cowpony was based on nothing except modesty. To have admitted that the horse had won sounded too much like bragging to Jack, for he was an essential part of each of the horse's victories. However, when Gibson told about this smooth black horse and the secrets about the victories were out, Jack rode over and roped Sailor so that we could photograph him. Later in the day he rode this black beauty and showed us the horse's marvelous shiftiness on foot. The pride of a fine cowboy in his excellent horse was never better demonstrated. Surely Sailor has few superiors or few equals anywhere on the Texas plains.

Personal modesty as shown by Spencer is a rather general trait among cowboys. As "Uncle Dick" Brannin expressed it, "If a cowboy brags on himself, nobody else will." Another trait noticeable among those who are in charge on the 6666 Ranch is the complete lack of eagerness to demonstrate authority. One might ride up to this crew at the wagon and study them for sometime before he discovered who was the wagon boss. The boys respect Jack Spencer because of his long experience and ability as a cowboy, and Jack does not have to lash out with authority to get things done.

After a rest period in which the boys lolled on bed rolls or otherwise made themselves at ease, they roped out their afternoon mounts and saddled them for the remainder of the day's work. Then they donned their chaps and rode back eastward across the highway to complete the day's round-up. In due time they returned to the branding pens with several hundred head of cattle—cows with their calves and an unusually large number of mavericks.

We had driven down the highway for a while and had not returned until most of the calves were branded. That part of the work was not so difficult. The cattle had been driven into the larger of the two pens and a few of these at a time were driven into the smaller pen where the actual branding work went on. In the branding pen itself Toar Piper threw a rope over a calf's neck or horns—if the animal was old enough to have horns—and

Riley Thacker roped the same animal by the hind feet. Then both men stretched their ropes placing the calf in a helpless condition while one or more of the cowboys "flanked" the calf, throwing him on the dirt on his right side. A mesquite wood fire kept the branding irons hot at one side of the pen and Jack Spencer handled the irons. He applied the big six four times high up on the calf's left side and with another iron Gibson stamped the date brand on the calf's jaw. If the calf was born in 1947 Gibson branded "7" on his jaw, but if a yearling (born in 1946) he burned "6" on the animal's jaw Accordingly, the date brand applied to the jaw of a two-year-old was "5." While the animal was down, Wayne Piper stuck the vaccination needle in each animal's neck, insuring against certain diseases.

"Business picked up" in the branding pen when the two-year-olds were branded. Piper and Thacker, who roped with precision and "put the stretch" on these big ones pretty nearly on schedule, "hubbed hell" with the big ones. It was hardly as little as a one-man job to lay a two-year-old bull on his right side. However, the day's work went ahead without mishaps.

These unbranded yearlings, two-year-olds and so on are known as mavericks. On a ranch like the Four Sixes they are simply a few that have been missed in previous round-ups. Somewhere near twenty of them—an unusually high number—were branded on this particular day.

All of the mavericks in the pen were finally branded—except one three-year old bull. He had his own ideas about this branding pen business. He pawed the earth and bawled and tried manfully to break out of the big pen. There would have been dramatic proceedings in the little pen had this mad bull been driven in for a taste of the hot irons, but better counsel prevailed and he was saved for the branding chute at a later date. Here he could be controlled and the brand applied without danger to the men.

It was as late as I could stay around these interesting scenes if I covered the territory mapped out for myself during the remainder of the day. The boys were about to kill a nice fat yearling which meant good fresh beef for supper. They invited us to stay, but despite the temptation to help consume the fresh tender beef we were forced to be on our way.

During the day I had fired a lot of questions at young Gibson, who of course, understood pretty well what the average American would like to know about cowboys. "What do they talk about around camp at night?" was one of those queries—and the answer was a complete surprise. Certainly you would suppose they talked about their dates and their girl friends in general. Of course, there was a little of that, so Gibson admitted—but what cowboy would pour out his real love affairs before a dozen or more men around a camp fire? Next came the supposition that one or two good story tellers in camp might hold the boys spellbound with yarns of an adventurous nature or even with a tall tale or two, but the suggestion again did not seem to strike at the center of the Four Six cowboys' campfire talk.

"The boys talk about their horses every night. That's almost all they do talk about." That was Gibson's answer to my battery of questions. The conversation does not run off so much on horses in general as on the particular horses in the Four Six remuda—"Jim" and "Blondie" and "Pippin Blue," "Junior," "Will Rogers," "Charleston Blues," "Pretty Boy," "High Power," "Darr Brown," "Eagle," "Friendly," "Sailor," "Old Churn," and at least a hundred others. Of course, the ranch owns all of these interesting cowponies, but the boys feel the full pride of ownership because each cowboy is assigned about ten of them for his exclusive use. Furthermore, if he chooses, he is allowed to trade one of his horses for a horse that is assigned to some other cowboy.

Here is the real spice of life on the Four Six Ranch. When the moon is tree-top high over the hills or the cedar brakes of King County and the camp fire has burned down to a bed of coals, the cowboys smoke and loll at ease across their bed rolls— and trade horses! The trades are not made by men who jockey to make a dollar by keen wits and chicanery. Here perhaps is one of the few spots on earth where men trade horses with all the mercenary motives removed. They trade for pure fun, trade for ease of gait or horse sense in an animal, trade for the love of a horse—these men who know and love cowhorses as well as anybody else on earth. I regretted not to have the privilege of hearing some of these novel horse trades.

In these visits to this giant Four Six Ranch at least one other thing of interest impressed me. That was the very human attitude of this multi-million dollar business toward its old cowboys. A cowboy who has spent twenty or thirty years of his life caring for cattle and battling the rough elements that are sometimes present in West Texas weather is not pitched out in his old age to shuffle for himself.

The case of Dick Brannin will serve to illustrate. Uncle Dick, as he became known, worked for the ranch a long time while Bud

Arnett was manager and continued his employment under other managers until 1947. For some fifteen years or more he has had charge of the Taylor camp. He with his wife, who was known to the cowboys of the ranch as "Aunt Alice," lived in this ranch house (Taylor Camp) and, if the chuck wagon was not at hand, provided meals and shelter for any of the men who needed to work in this southeast part of the range. Uncle Dick made a hand when the wagon was nearby, and he looked after the ranch property in his area at all times.

For a long time he had made a good hand—ever faithful to the best interests of the Four Six Ranch. Then in time the white hairs began to show up in Uncle Dick's head, and at length a slight rheumatic condition made it pretty difficult for him to throw his leg over the saddle when he mounted his horse each morning.

Aunt Alice began to notice Dick's heroic efforts and prayed silently for some kind of relief. Finally relief came in a form that she had least expected. George Humphreys in company with two high officials of the ranch rode out to Taylor Camp one day after Uncle Dick. They had him go with them, presumably to help with some ranch work, but actually to tell him about a decision that concerned him. They had decided to retire him on full pay for the rest of his life. Uncle Dick took the sudden shock heroically, just as he had done his work for some time under his physical handicap. He agreed to the retirement, but as was characteristic with him he said very little.

When he returned to Taylor Camp, Aunt Alice could tell by his actions that something unusual had transpired. She asked Dick what it was, but the hardened old cowhand overcome with emotion—part joy, part sorrow—could not find words to express himself. Speechless and with a big tear drop at the corner of each eye he hesitated for a while before he could regain his poise. Aunt Alice through intuition somehow knew that the news which Uncle Dick would finally tell her would not be bad. She waited and then finally listened when he told about his retirement. She was so happy at the thought that he could at last have relief from duties that taxed him a little beyond his physical capacity—and yet she was sad that the day had finally come when he could no longer carry an even load with the younger cowboys.

Soon they moved from Taylor Camp to their new home, a mile south of Benjamin. It was here that Ernest Lee and I visited the Brannins in their recently constructed concrete block cottage—neat, clean and white—just as one could have known that Aunt Alice would have planned it.

We picked up another thread of the Four Six Ranch story in the course of this stay in King County. The wild horse that Boog Graves captured some eighteen years earlier made the Four Six Ranch a good cow pony. In the remuda and to the cowboys he had become known as Nester Boy and his reputation for usefulness had continued to grow through the years. Then Nester Boy, like the famous old Hub, grew too old to do a full share of cow work and was about to be replaced by younger horse flesh when Mart Robinson from Paducah, Texas, came along one day and offered to buy him. George Humphreys, ranch manager, was not tempted by the small sum of money offered for the old horse, but he knew Robinson and trusted him as a gentleman.

"Mart, I'll sell him to you for that price if you'll give him a good home and promise not to sell him," was George Humphreys' answer to Mart's offer. Mart agreed and the old horse was moved to the Robinson farm in northwest King County. It was here in his new farm home that Lee and I found Nester boy, the one time wild horse that made a faithful Four Six cowpony. He was sleek and fat in testimony of the faithful manner in which Mart Robinson has kept his word, but the incident reminds one even more forcefully of the fine retirement policy of the Four Six Ranch. Here is one big business institution that does not wring the last drop of profit out of either its aging men or its old horses.

The Burnetts who maintain this humane retirement policy were previously a part of several dramatic episodes down in Wichita County. We shall return to their story when this journey turns back to the country around Wichita Falls and the town of Burkburnett.

Going back to King County and the 6666 Ranch let us resume our narrative. I drove to Guthrie where Ernest Lee left me. Turning again southward, I continued this extended journey through the ranch country alone—but not alone either for there were interesting people not many miles ahead.

4—Long Whiskers and Riches

The car glided along smoothly over the paved road toward Aspermont. Near the half-way point of this leg of journey, Kiowa Peak mentioned in an earlier chapter was faintly visible some fifteen miles to the east. Long before oil excitement was known in West Texas, this small mountain was the scene of a mining bubble.[1] Here it was that unsuspecting investors spent their money with some promoters whose imaginations had learned to roam in green pastures. There were great quantities of copper bearing deposits in and near the little peak, but most of it was low grade in quality and the cost of producing and transporting the mineral to market was far greater than the actual value of the copper when sold. The mining venture ran its course quickly. By 1879, Mark Lynn of Palo Pinto, Texas, moved his L I L brand[2] of cows out near this peak; and copper mining gave way to milling cattle, jingling spurs and broad brimmed hats.

Soon my road dipped down into the rough country beside the Brazos. An expert with a camera can capture some beautiful

1. Capt. R. G. Carter, On the Border with Mackenzie, Eynon Printing Company, Inc., Washington, D. C., 1935, pp. 332-36.
2. See George B. Loving, The Stock Manual, 54.

scenes from the roadside here. At one point where the highway curves around a hill top, the Brazos Valley with its myriad little hills and lush greenery spreads out like a giant map—a map of untamed country that is most likely to produce beef-steaks for yet other generations of Americans.

Shortly the road passed an area of sand dunes and entered the smooth country again. Smaller subdivisions of land were evident. Again there were farms with little herds of cattle scattered here and there. A spring stream called Stinking Creek lay just ahead. The name was not an accident; it was handed down from the days of the great Buffalo slaughter in West Texas. The creek with its unpleasant name was in the heart of this greatest game hunt of all time.

In Aspermont I met John Guest who came to the area in 1886, a date known to all pioneers as "the year of the drouth." Guest stayed and prospered in this land of uncertain rainfall for the full six decades that followed that year of adversity, but neither drouth nor the "gyp water" for which the country is noted have hindered his progress. He still remembers the old road three miles south of Aspermont along which west bound emigrants in covered wagons were not uncommon sights sixty years ago. The road was the first thoroughfare made and used by white men traveling to the Lubbock-Plainview part of the high plains. General R. S. Mackenzie opened the road on a military expedition in 1871.[3] Since that date, the old trail originally made by military wagons has been traveled by buffalo hunters, hide haulers, freight wagons, cattlemen and farmers—all in the westward push of civilization that has builded modern cities on top of the Plains which for many years was believed to be a great American desert. The old trail has long since ceased to function; wire fences chopped it into 100 little pieces at least a half century ago, and wind and rain have all but erased it from the landscape. Guest pointed out two faint traces of this earlyday highway in his pasture near Aspermont. Beside the old road at one point now stands a giant mesquite tree knotted and gnarled by the storms of many winters. It was there sixty-eight years ago when the Guest family first came to Stonewall County. One can not refrain from comparing the old tree with the pioneer Texan who has remembered it all through the years. The two have been partners in the development of the West.

3. Capt. R. G. Carter, **On the Border with Mackenzie**, 149-217.

The old car played an embarrassing trick on me right in the middle of the main thoroughfare in the business section of Aspermont. Heat from the motor caused by a new repair job vaporized the gasoline in the fuel pump when I stopped the car for a moment. The engine when started again, ran only long enough to exhaust the fuel that had not been vaporized, and then it stopped in the middle of the street like a balky old mule. It was in front of a drug store—and for the first time in my life I asked a druggist instead of a mechanic to assist with my car troubles. His prescription was simple. He handed me a glass of water and suggested that I pour it on the fuel pump to cool the vaporized gasoline—and believe it or not he was one druggist that knew something besides rolling pills! The prescription worked like magic and shortly the old motor was back at work again as good as new. So successful was the remedy that I bought a gallon bottle and filled it with water to insure against similar emergencies in the future.

From Aspermont the journey turned west. The pavement led across Stinking Creek again and into the area of shinnery and deep sand that covers no little of western Stonewall County. Bright pink and orange colored flowers lined the roadside from time to time—a botanist would have been a great help at this point, for my knowledge of wild flowers is next to nothing. A quail jumped up and ran across the road just as if he realized that the game laws gave him full protection. Actually he had not the least cause for worry, for my shooting eye is none too good—and, well, I didn't have a gun anyhow.

Ten miles south of my road the Double Mountains stood out in plain relief. These two great mounds that can be seen for forty miles in some directions have been a landmark to travelers since the days of the California Gold Rush.

The highway soon crossed the Salt Fork of the Brazos and entered a land given over to farming. Occasionally the pink and orange flowers again showed up by the road side, but the quail was not anywhere in evidence. Perhaps it was too far from the shinnery for quail to venture.

Late in the afternoon I entered the little town of Jayton, which serves the nearby farm belt. Sixty years ago both town and countryside presented a far different picture from the scene of today. Wild open country then, it was well within the range of the Two Circle Bar Ranch, O. J. Wiren, the owner of the old ranch, was a long bearded Scotchman who had not shaved in years. Curiosity ran rife among the cowboys as to the cause of his unclipped beard. One day a nearby cattleman inquired as to the cause of his objection to use a razor. The old Scotchman explained how he had prospered in the cattle business and had amassed a fortune that ran well into the hundreds of thousands of dollars. "I have promised myself," explained Wiren, "not to shave until those whiskers cover the face of a millionaire."

So far his whiskers and grass and cows had grown in profusion, but there were troubled times ahead. At least a part of the cattle were moved to the "Strip" in what is now Western Oklahoma where they could expand apparently without limit. Soon, however, the United States Government ordered the cattlemen to leave that part of the range. Actually the grazing land belonged to some tribes of semi-civilized Indians. The order caught Wiren in a difficult position. He had to work out his grass requirements at a time when many small cattlemen were crowding the range. The consequent loss greatly reduced his fortune,[4] and so far as anybody knows, the long beard worn by this old Scotchman never did cover the face of a millionaire.

Wiren had orginally acquired his Two Circle Bar cattle from J. J. Hittson of Palo Pinto County, a county that produced many of the successful cattlemen of the late 1870s. Hittson had started the brand in 1874 moving the herds with this peculiar mark to the Double Mountain River about five years later.[5] Soon the bearded Scotchman bought them all and made them the basis of his short lived fortune. By 1886 some 30,000 cows, perhaps, proudly wore his Two Circle Bar brand on an immense range of some 500 or more square miles of grass. So great was his dominion that sunrise caught some of his herds nipping grass in the shade of the Double Mountains in Stonewall County while sunset found others grazing contentedly far over in Kent County, a full day's buggy ride to the west. Some of his cows sipped the putrid waters of Crotan Creek several miles to the north of present Jayton, while others found their liquid refreshments in small water holes on the south slope of the Double Mountain River thirty miles away. Truly it was a veritable kingdom but Wiren, its hairy faced king, was shortly to lose his throne.

4. The story was told to me by the late John Bryan of Abilene, Texas.
5. Informant, the late J. T. Bond of Jayton, Texas. The Two Circle Bar cattle were on the Double Mountain River by 1881 when Loving's brand book was published. J. J. Hittson who still owned the brand at that date posted this warning in his advertisement: "Will pay $100 for proof of any person using any of my cattle without authority from me." See George B. Loving, **The Stock Manual**, 56.

Ownership passed to Kellog, McKay and Runnery in 1889, but the Scotchman's old brand with a slight change, like an ancient coat of arms, was destined to last on through the years.[6]   During the next decade men by the names of Clark and Cravin became owners in succession—but neither of them made any change in the brand.   Then in 1901 new owners came to the helm—perhaps it is not stretching the figure too much to say that a new dynasty of kings came to the throne.   John A. Wishard and John S. Bilby became owners and the brand was changed so that the bar was placed between the two circles (one circle above bar, one below) instead of two circles side by side with a bar below, as Hittson and Wiren had run it.   For yet another decade or two herds bearing the changed insignia still approximating 15,000 cattle were scattered over much of the old Scotchman's original range—in Stonewall, Fisher and Kent counties.   Finally about thirty years ago, a sizeable remnant of the ranch became the property of Bert Wallace, a grandson of the former owner John S. Bilby.   The brand now called "O Bar O" instead of Two Circle Bar has not suffered any new changes—and lest the reader should feel that the old Scotchman's empire had gone to pot, one may still ride twenty-five miles in a straight line across both the Salt Fork and Double Mountain rivers without leaving the lands of Bert Wallace.   The Two Circle Bar brand, now called the O Bar O, as found on much more than 100 square miles of Kent County land is older than any human habitation on the plains of Texas north of the Texas Pacific Railway and West of the 100th meridian.

In Jayton as in most of the places along my way I met "head on" the genuine West Texas kind of courtesy.   In some quarters one meets with the type of courtesy that smiles and bows politely out of the picture for fear of wasting a few moments of supposedly valuable time. At other points one encounters the kind of courteous person who has something to sell and is courteous only so long as his actions promise to pay cash dividends.   The type of courtesy encountered at Jayton was far more genuine than either of these. Everybody offered to assist in obtaining local information. Bill Harrison gave some two or three hours of his time to insure the success of my local visit.

He led the way to the home of J. T. Bond, an old Two Circle Bar cowboy. Bond had been in the nearby plains country for sixty-five years and had been near the place that is now Jayton

6. This account of the Two Circle Bar cattle and their owners is from the late J. T. Bond of Jayton, Texas. Also see Gus L. Ford, **Texas Cattle Brands**, 86.

since 1886 when he first rode herd for the above mentioned Scotch cattleman, O. J. Wiren. Sandstorms and drouth had not discouraged him; on the contrary he, like John Guest, had prospered in spite of the periodic ill temper of the elements.

Bond came west in 1881 when the great grass range, recently hunted clear of buffalo, was a paradise for the little cattleman. He had watched the growth of the big ranches financed by eastern or Scotch capital and had noted the continued influx of the smaller herdsmen until the range was overcrowded. It was in 1885 that he remembered meeting ten wagon loads of barbed wire in transit from Colorado City to the giant Matador Ranch—because the day had arrived when fences alone could save the grass from the mouths of too many cattle.

Until 9:30 at night Bill Harrison and I traveled the trails of the old west with this sturdy pioneer, and then Bill telephoned to arrange for a room at the local hotel. It was not in a cowboy's bed roll but under a friendly roof that I slept on the range of the old Two Circle Bar.

At perhaps 4:00 o'clock in the morning the chickens crowed where in time past one would have heard the low of Two Circle Bar cattle. I lay gazing out the window looking across the landscape where the cowmen—small, medium, and large—had come to take over the west only a little more than sixty years ago. Jesse Hittson, the Slaughters, the Daltons from Palo Pinto county; Hensley, Crill, Crutchfield and the Coopers from Jack county; George Edwards, Dan Cole and (in a small way) Hank Smith from Fort Griffin and Captain Jim Reed from Fort Worth and a list of some length from many points in central Texas—a veritable little army of them had come up and scattered over a vast range on the Upper Brazos from a little west of the old town of Fort Griffin to the Caprock of the Plains.[7]

Hundreds of buffalo hunters had been there in 1877 for their last big hunt—the hunt of extermination—for the buffalo never again came back in great herds and the hide collecting business was near its end. Thus the range was left clear and the cattlemen in bordering counties were not slow to make the most of their new opportunity. By the end of 1879 the grassland in the valleys of the Salt Fork and the Double Mountain rivers were grazed by so many herds of cattle that a regular round-up organization became necessary.

7. The **Fort Griffin Echo** files for 1879 and 1880 covered well the news of this area, listing many more references to cattlemen than can be published here.

Early in February 1880 these new conquerers of the plains
met on Chimney Creek, a small stream in northeast Garza county,
and organized the Blanco Canyon Stock Association. It was by
no means a midget on the day of its birth; twenty-five cow outfits
were represented and a regular program of work was mapped out.
On the open range it was necessary from time to time to round up
the cattle from all over the country and send back the strays to
each of the several ranches from which they had wandered.   A
round-up boss was appointed for the whole upper Brazos Valley;
chuck wagons and supplies were furnished somewhat in proportion to the size of each of the ranches involved; and all of the sub-
scribing cattlemen, large and small, were to be represented directly or indirectly by one or more cowboys.

Except that the purpose was entirely different, this regional
round-up was like a giant rabbit drive. Ranch at a time was
combed for cattle in this grand scale systematic procedure to find
and collect the strays and get them back to their proper range.
At the appointed time the coyboys sent out by these men of the
Upper Brazos met in Blanco Canyon some fifty miles east of present Lubbock and began their work. They worked westward and
southwestward in Crosby and Garza counties, then turned eastward down the Double Mountain River reaching perhaps as far
as modern Rotan, and finally they worked northward and northwestward up the Valley of the Salt Fork past the places where
the towns of Jayton and Spur have since established themselves
as trading centers.

The old *Fort Griffin Echo,* one of the top-notch frontier newspapers of its day, told the story of the meeting of cowmen on
Chimney Creek almost as fresh as one might find in yesterday's
newspaper.[8] The names of all of the ranches in the exact order
in which the big cow hunt was to proceed are given in this seventy-four year old newspaper probably without any omissions
even though there may be errors in spelling. It is doubtful if any
of the cowboys who participated in this first big round-up of the
Upper Brazos are still alive; but since the list of ranches may contain names that will call back memories in the minds of many of
the readers, they are all included here. The round-up began on
Wolf & Parsley's ranch in Blanco Canyon then moved to Shank-
lin's and next to Slaughter Brothers and from there to Kyles, to
Dalton's to Hensley and Crutchfields, to Kidwell and Runnels,

---

8. **Fort Griffin Echo**, Feb. 7, 1880.

Crill & Fabriques, Gailbrath & Hay's, Powers and Beal's, Cooper's, Lindsey's, Segres, Elkins & Jackson's, Hittson's and to Nuns. These last ranches were southward of the Double Mountains and some sixty or seventy miles from the place where the round-up began, but there was yet work to do. The return trip worked the ranches of Gholson & Taylor, Purcell, Gambill, Gaffe & Weir, Pepper & Browning, Coggin & Wiley, Campbell, Hall and Baker.

Such an organization as this Blanco Canyon Stock Association seems strange when one remembers that these men of the West—men of the extreme Texas frontier—were some of the most pronounced individualists that inhabited this planet. They were men who hated regimentation with a passion and did not propose to give up any of their freedom of movement until it became necessary. But after all this is said, these were practical business men who recognized the advantage of co-operation. There was not a fenced ranch in West Texas, and the only available substitute for fences was this method of systematic combing of the prairies to hunt up the strays.

But there were annoyances other than the fact that cattle slipped away from their owners. A few prowling Indians still occasionally visited the plains and worse than that, this new country had its quota of cattle thieves. Just as early Texans had organized companies of rangers or as colonial Americans had formed units of minute men, these men of Blanco Canyon set up their own organization for frontier defense. The number of guns and the amount of provisions that each ranch (somewhat according to size) was to furnish was agreed upon and R. B. Segres was made captain with W. B. Slaughter as his lieutenant. These cow men of the plains, though individualists in the extreme, knew how to set up the essentials of local self-government and especially of self-protection whether they proceeded according to the law books or not.

The military feature was perhaps unique as a function of a cattlemen's organization, but the regional roundup had already become standard procedure for the open range country. The cattlemen of what one now thinks of as the Abilene country had met at Buffalo Gap and perfected such an organization.[9] The cattlemen of the Panhandle were soon to meet (in 1880) at Mobeetie and organize.[10] Probably a little later a regional roundup was

9. **Fort Griffin Echo,** April 26, 1879. Reference is made to a meeting of the Cattlemen at Buffalo Gap.
10. **Fort Griffin Echo,** August 7, 1880.

set up in the upper Wichita and Pease river valleys. But the original of all cattlemen's organizations of the kind met at Graham in February, 1877.[11]   That it has since grown and swallowed up all the little ones is a fact too well known to need elaboration.[12] Men who engaged in the cow business knew how to meet their problems whether it was the return of stray cattle, the punishment of thieves, or the tracking down of wild Indians. The pioneers of the Upper Brazos ran true to the usual formula and added a little of their own medicine to insure its potency.

11. Graham Leader. March 9, 1877.
12. The cattlemen of the upper Brazos continued their roundup association through the year 1887 after which the great amount of wire fencing made the association unnecessary. See W. C. Holden, Rollie Burns, 136-37.

5—The Prince of Good Neighbors

With these many interesting details of early days of the upper Brazos River country running through my mind, I dropped off to sleep. I was still in my hotel room in the little town of Jayton. Somehow I felt the urge to get out of bed and point the nose of the car due west into the heart of this upper Brazos country where these pioneer cattlemen of the plains held their great cow hunt back in 1880. By seven o'clock my face was separated from its coat of whiskers and I had started downstairs ready to begin the day. After a little food and a check up to see that none of my notes were left behind, the trek began down the road southwest toward the Salt Fork of Brazos. Some six miles of Jayton I crossed the path of the old MacKenzie Trail—there was no sign of it from the roadside although the land maps at Austin have preserved its location here so that future generations can know its exact course. It was slightly more than a mile to the Salt Fork where a great belt of the ranching country began in earnest. The above mentioned ranch of Bert Wallace, modern successor to the old Two Circle Bar, began at the river bank and extended along most of the half dozen miles of highway that led to the tiny county seat town of Clairmont. This remnant of the land once ranched by our bearded Scotchman of the 1880's extended ten miles north of the road and fifteen miles south of it.

Clairmont, according to my road map has 175 inhabitants, but I saw only about a half dozen of them and wonder where the census taker found the others. The most impressive of the few houses in this little town was its tiny court house—two stories high and made of red sandstone. It was certainly one of the smallest to be found in Texas, but probably large enough to meet the requirements of the sparsely populated county.

Across the street from the court house was the filling station owned by "Red Mule Barkley," an ex-cowhand of fifty years ago. A half century ago he was horse wrangler for the Two Buckle Ranch which was located in Blanco Canyon near present-day Crosbyton about fifty miles northwest of the site of Clairmont. The cowboys of the ranch gave Barkley his nickname which has stayed with him all through the years.

Red Mule called my attention to the sizeable property of the Morgan Jones heirs—fifty-three square miles of pasture land that lay just west of both the Bert Wallace Ranch and the little town of Clairmont.

He also knew some of the first settlers of this part of Kent County and in particular, remembered Boley Brown—the same fine sportsman who was runner up to Sam Graves and old Hub in their famous contest at the Cowboy Reunion at Haskell. As a boy Brown had come to this part of the cattle country soon after 1880. He was an orphan with little more than an alert mind and a will to work. In the face of a hard blowing sand storm, he arrived in the upper Brazos country with a herd of cattle—a cowboy much weather beaten and doubtless not a little discouraged. But he stayed and grew with the country. Soon in this land of open range and free grass he began to build up a herd of his own. Before 1900, he had become a big cattleman. Far and wide, everybody had come to know the name of Boley Brown. As a partner of Pete Scoggins, his "24" brand covered thousands of acres in western Kent County. Brown was the kind of a fellow who never had a press agent, but now forty years after his death his neighbors remember many little stories of his philanthropic nature. Shortly we shall return to his story.

While Barkley told me about these several early day cattlemen, the residents of this little western town were beginning to busy themselves with their daily chores, and it was perhaps a good time for my journey to continue. My highway led west through the heart of the Morgan Jones Ranch. A. C. Cairns, an Englishman, acquired the property more than four decades ago and held it until it was sold to the present owners. Some four or five miles west of Clairmont and to the left of the road stood the ranch headquarters of the Morgan Jones heirs. A few minutes spent at the place, camera in hand, netted at least some record shots as well as a memory graph of this part of Texas that had formed only slight acquaintance with the plow.

But all these millions of acres and their grass did not intrigue me as did Red Mule Barkley's story of the princely cowman who had become a tradition in this land below the Caprock. The spirit of Boley Brown still seemed to hover over these cowboys and these herds of cattle like the rich yellow glow of today's twilight. It was the country across the highway from the Jones headquarters that had been the center of Brown's world while he had grown from a modest cowhand to a man of importance—during the years when the West had changed from open range to fence-rimmed pastures. He had lived and died out there a few miles across the modern little ribbon of concrete over which today's world hurries by. His old home was not far to the northwest.

I drove out across country to Boley's home and tried, under the stimulus of his own surroundings, to understand why the old cowpokes of his area still praise him forty years after his death. Nobody was present to aid in my quest for enlightenment. Soon I drove on through the pastures to another part of the range— then all of a sudden it occurred to me what Red Mule Barkley had said. "Boley Brown was the kind of fellow who would catch a poor old nester up against it and pay him twenty dollars for an old ten dollar cow. More than that, he'd make the nester feel like he thought the cow was worth twenty dollars." Barkley had already answered my question although it had taken a little time for me to realize it. In his brief statement he had told of Brown's unselfish neighborliness and of his charitable heart that made every man his equal.

I had just parked the car at a point on the old abandoned road that at one time, had led northwest from Clairmont. It was here, back in 1911, that Boley Brown had slumped down in his buggy and died of heart failure. His body was removed to a nearby county for burial, but somebody—I never learned whether it was relatives or friends—erected a monument to his memory here on the range where he had lived and died. Even today, this clean cut shaft of stone is fenced in a little enclosure and neatly kept. So far as I know, it is unique in all the West. Not for any deeds of heroism but just for neighborliness and fair dealing, the memory of a deceased cattleman is kept alive.

One may drive into Clairmont or go far and wide into adjoining country and still find old cowpokes who will leave no doubt in anybody's mind that Boley Brown was the prince of cowmen. Here on the range where he lived among his fellowmen somebody—without using the words—said the same thing with a monument of stone.

It was only a few miles along ranch roads back to the paved highway. Once out on the open road, I picked up the thread of my journey where I had left it.

In traveling westward through this land of old ranches, I was passing over the ground on which the cattlemen who had organized at Chimney Creek in 1880 conducted their extended roundup. Though the country had gone a long way from the free grass and open range of 1880, it was still ranch country, only now it was under fence and was sliced in halves, as it were, with a good concrete road. There was only one plot of cultivated land in sight of the road for nearly forty miles west of Clairmont. Twelve miles west of the little town the highway passes up onto the divide between the Salt Fork and the Double Mountain Fork of the Brazos River. The valley of the Double Mountain River yawned in the morning sunlight as I gazed at Mount Mackenzie fifteen miles to the south. This prominent peak was named for Colonel Mackenzie (later General Mackenzie) who opened a wagon road beside it in 1872. The road was Mackenzie's approach to the high plains on one of his several Indian campaigns.

Shortly I reached Garza County and somewhere a little farther along the way entered the old O. S. range of Andy and Frank Long. Like many of the other early ranches the area is now divided into various pastures and ownerships, but the old original O. S. brand is still maintained by one of these several ranches.

Two quail ran across the highway soon after I entered Garza County. There was little else to break the monotony except that one might muse as he drove along this fine smooth highway as to the hardihood of the Slaughters, the Daltons, and others of the first pioneers who settled this upper Brazos Valley. Not everything they did had survived the intervening seven decades that had elapsed, but much of the true flavor of the old West began in their day.

Many of their old brands are gone now. Yet here as I gazed across limitless grasslands now fenced with barbed wire, somewhere in the adjoining pastures were whitefaced cattle still bearing the O. S. brand that once belonged to Andy and Frank Long.

The story of this brand is somewhat similar to that of the Two Circle Bar and many others. As Cinderella rose from kitchen maid to great lady at the ball and then reverted back to rags, so did many of the old West Texas brands—except that the reversion back to rags is not at all a true parallel. During 1879 and for several years afterwards, little individual cattlemen came from the twenty or thirty counties that form a long north-south belt west of Fort Worth to the foot of the great plains. Most of them numbered their herds only in hundreds; a few could count their cattle in four figures. Then rather suddenly because of good management, greater capital or credit, a few of these individuals began to build up great ranches, to buy out the little fellows and to establish grassland empires. Also in some instances promoters built these inland principalities with eastern or Scotch capital until ere long several dozen brands covered much of the map of the West Texas cow country.

Up to this point the quick growth of the big ranches has been much like the rapid rise of Cinderella at the ball, but the close parallel stops there. With Cinderella the hour of triumph ended when the clock struck twelve. With the great ranches their day of prominence did not end with the sounding of a gong—in fact many of them have not yet reached a point of decline. However, perhaps the greater part of them have begun to give way under the pressure of an irresistible force. Not like the quick striking of the clock at midnight but more like the gradual beginning of a day or perhaps more like the slow dawning of an age, many of these great cattle empires have yielded to the relentless pressure of this irresistible force.

What is the force that has thus continued to hammer away at the cattle kingdoms of the west? Not the heavy pounding of a sledge hammer against an anvil as the question seems to infer, but the slow plodding of ponies' feet and the creaking of wagon wheels. It was a west bound stream of covered wagons that had come out of the East with the intent of invading the great plains. These wagons were occupied by farmers with their plow shares, pots, and pans and were followed by their dogs and milch cows. To the cattlemen these intruders were known as nesters, a term that carried with it nothing in the nature of a compliment. The ranchmen tried to discourage this stream of immigration. In instances they intimidated the newcomers and at time held them in contempt, but the nesters continued to come. First their covered wagons came by dozens, then by hundreds, and finally by thousands. They planted cotton, corn, sorghum, and other crops until after a series of pioneering efforts involving some of the most widespread experiments known to agriculture, they found it possible to live on these semi-arid prairies. Numerically these farmers from the East soon dwarfed by comparison the little group of cattlemen and cowboys that had first spread themselves across West Texas. When counted throughout the whole of the great Plains, they made even the California gold rush seem like a small migration.

As their numbers continued to swell, land values began to rise and although there was sometimes bad blood between the ranchers and nesters, these little men from the East were often the salvation of a rancher who had run into financial difficulties. It was true that no one nester could aid the big cattleman financially, but the presence of this new farming class in great numbers made his land salable at a profit even if he had had reverses in his management of cattle. Land advanced from a dollar an acre or less to two dollars, then three dollars, then five dollars, ten dollars, twenty-five dollars, and in many instances more than twice that amount as the great horde of farmers came west between 1900 and 1920. The owner of a 200,000 acre ranch had become a multi-millionaire! Land that was adaptable to farming was in many instances considered too valuable for grazing—and big ranchers sold out to farmers. The nester and the market for land that came with him were the irresistible force that was to melt away many of the great ranches! The change did not occur as the clock struck twelve or any other hour of the day or night, but slowly and in varying degrees at different places it remade the real estate map of the West.

The story of the O. S. brand is not a perfect example of this West Texas ranch cycle but it has some features of the general rule. In 1878 Overall and Sharp[1] who appear to have lived at Coleman, Texas, started this brand formed from the initial letters of each of their names. By 1881 Sharp must have sold his interests to Overall for the faded leaves of the Stock Manual of that date lists the O. S. brand as the sole property of an R. H. Overall[2] of Coleman, Texas. But according to the old brand book, the cattle had already been moved to ranch lands at the "head of the Double Mountain Fork of Garza County." The brand was

1. Gus L. Ford, **Texas Cattle Brands**, 150.
2. George B. Loving, **The Stock Manual**, 190.

new to the upper Brazos valley, but it was destined to live in the new country for a long time after most of the other brands represented at the famous meeting on Chimney Creek had been forgotten.

In 1882 this now famous brand was sold to Andy and Frank Long, whose ranching interests spread over perhaps a quarter of Garza County. The Longs and their herds held forth for twenty years—and then the property was sold to Connell, Clark and Scharbauer. Connell and Clark acquired the holdings in 1914 but the old brand was never changed.[3] After another half dozen years the Connell brothers became the sole owners of this same O. S. brand. Perhaps they and their descendants who have survived them have held a certain inner sense of pleasure from the ownership of this famous old brand somewhat similar to the sensations experienced by a collector of fine old paintings.

3. For a brief sketch of the history of the O. S. brand, see Gus L. Ford, **Texas Cattle Brands,** 150.

6—Post Toasties and Plowed Ground

The details that have to do with the growth cycle of the O. S. Ranch are omitted here because another example lies only a few miles down the road, an example that involves not only the cattle days of the early West but connects them with a most interesting venture of a multi-millionaire of Battle Creek, Michigan. It ties together in the story of a single ranch, both cowboys and nesters, and salts the mixture with cattlemen of magnitude and also includes a big time Eastern business man whose breakfast food is better known than any cattle brand of the old West.

I drove down the pavement past this present day O. S. Ranch into the spic-and-span up-and-coming west Texas town of Post City. At the end of my forty mile ride of frontier solitude, I went into a very modern drug store to purchase a cool, refreshing drink. Here I sat in modern urbanity that by a strange shifting of events had come out of the old Currycomb Ranch of other days. There was a cotton mill in Post City that produced fine Garza sheets and pillow cases on the same large tract of land where many an early day cowboy had gone to sleep in his bed roll and perhaps dreamed of milling Currycomb and Y. G. cattle. The cotton mill and its reserve space, of course, covered only a few acres of ground, whereas the ranch at its best covered 200 square miles of grasslands. It was roughly the northwest quarter of Garza County and included the site of present day Post City. Yellowhouse Canyon in its southeastward course cut across the big ranch leaving most of the land southwest of the Canyon. Yellowhouse Creek at the bottom of the Canyon is the principal tributary of the Double Mountain River. It joins the latter stream about thirty miles southeast of Post, but it rises more than 150 miles to the northwest. Yellowhouse Canyon through which this long creek flows is a mile or two in width at Lubbock, but it widens out to a much broader valley in Garza County. Post City is in this broad part of the Canyon a few miles west of the Creek. Thus the former ranch lay partly in the canyon and partly on the flat plains to the southwest of it.

It was up this canyon some ten miles north of the place where Post City is located that Young and Galbraith drove their small herd of cattle with the Y. G. brand in 1879 . Three years earlier they had begun near Fort Worth with only 300 head of cattle. They had grown some in the intervening time but their big expansion was yet to come. In 1880, the very same year that the cattlemen held their meeting on Chimney Creek, these two partners increased their business to a $400,000 capitalization. The Llano Live Stock Company was the new name given the expanded ranching venture. Sam Gohlson, a neighboring cowman, traded in his 2500 head of cattle for a $50,000 block of the stock and a number of Dallas and Fort Worth men became stockholders. In time the big tract of land in northwest Garza County was purchased and the little Y. G. Brand that began near Fort Worth had become one of the large ranches. The herds multiplied and soon reached some 10,000 cattle.[1] However, the Y. G. brand that Young and Galbraith ran at first, was superseded by the Curry Comb soon after they came to the Yellowhouse Canyon.

The Llano Company continued to do business until 1907 when it had an opportunity to sell its lands to a unique character who would have aroused colorful comments by the cowboys had they known he was about to buy a ranch. The strange investor was none other than W. G. Post, the Post Toastie king of Battle Creek, Michigan. He was not actually in the market for a ranch; what he wanted was land. He had his own ideas about the possibilities of colonizing in West Texas, and ignoring the jests of those who knew about southwestern drouths, he turned the venture into an outstanding success.

It was not long until he had sold tracts of land to a hundred farmers who were given a long term payment plan at an attractive rate of interest. There were more farmers to come and Post built

1. W. C. Holden, **Rollie Burns**, 106-7. The story of the Llano Live Stock Co., is told in some detail in a whole chapter of this excellent account of a pioneer cowboy.

them a trade and social center in the village of Post City that offered many features for other West Texas towns to envy or imitate. The many unusual things about Post City and the new colony are beside the mark here. The fact that interests us here is the dissolution of the one time great Currycomb Ranch and the building up of a compact community in its place, in which people were almost as numerous as Currycomb cows had been. Garza County had less than 200 inhabitants in 1900, but in 1910, three years after Post began colonizing, it had nearly 2000; and by 1920, nine years after a railroad had come to Post City, this figure was doubled.[2] Another empire in the land of cattle had melted before the plow. Will this continued breaking up of the large ranches eventually take your beefsteak from the dining room table and force the use of less palatable substitutes? Hold your fire until a more complete picture of the entire ranching story is at hand and the answer can come with greater accuracy.

After sipping the last drop of my cold drink, I left the drug store and drove out of Post northwest up the highway toward Lubbock. Three or four miles on my way I motored up a long steep incline to the top of the great flat Plains of West Texas. I was now above the Caprock. At a glance it was evident that I had left the ranching country behind me. Back toward the east and northeast lay the great broad canyon or valley of the Yellowhouse. From my vantage point two or three hundred feet above, I could see many miles of this long uncultivated ranching area that I had traversed on my way to Post less than an hour ago. Raw and broken or rolling country, it presented much the same appearance as when viewed by the first white men. Not so the level prairie to the west and southwest. In that direction as far as eye could see were farms, farms, farms—the dream of the Post Toastie king had come true. Dozens of these well cultivated parcels of the former Currycomb Ranch were in plain view. Each one of them had its windmill and with few exceptions its grove of cottonwood trees. Each of them across the stretch of level acres seemed to float in a lake of mirage.

Here, almost unbroken for a hundred miles to the west and much farther than that to the north and northwest, were the high flat level plains that had inspired an emotion akin to fear in most explorers. Men had died of thirst in those monotonous millions

2. The Texas Almanac, 1949-50, 102.

of acres.[3] Some had become crazed as they pursued mirage after mirage, hoping against hope to find water. At last civilization had spread across this broad expanse of plains and had made it a garden spot. Given enough rainfall or irrigation it can produce like a fertile river valley.

Much of the 200 miles that I had driven before reaching this high table land was not adaptable to farming. It had been the great grazing belt across which some ninety percent of the Texas buffalo had lived from time immemorial. These great shaggy beasts had tramped northward and southward up and down this strip of grassland as far back as our knowledge extends. A smaller remnant of these native quadrupeds had lived on top of this high level Plains country. Even a buffalo must have water and perhaps some shelter from blizzards.

Similar reasons have made the High Plains less adaptable to ranching than the broken country that borders them on the east. This rather wide fringe of rolling prairies that lies below the Caprock along my road from Wichita Falls to Post City, once the paradise of the North American bison, has become one of the great ranching belts of the West. One can mount a horse at the east edge of the Waggoner Ranch some thirty miles from Wichita Falls and, except for a small skip of less than ten miles, ride 150 miles across nothing but ranches all the way to the edge of the Caprock near this town of Post. In making this long horseback ride, one would cross all the large present-day ranches mentioned in this story so far and the Masterson and Pitchfork ranches besides, to say nothing of several smaller pastures.

The comparison between this belt and grazing land and the High Plains calls to mind the argument that took place between an artist that painted a cover for the *Saturday Evening Post* and an old Amarillo cowboy.[4] The artist looked at the level prairie about the old chuck wagon which he was about to sketch and remarked that the flat country would not do for range in his picture. The cowboy wondered what would serve as range if not that. Actually not all of the argument rested on the side of the cowboy. To be sure the particular tract in question had fed cattle for a long time, but traditionally both the favored range of the

3. In 1877 a detachment of soldiers and a party of buffalo hunters met with terrible privation on these Plains. Some of them died of thirst. John R. Cook, **The Border and the Buffalo**, 265-285.

4. **Saturday Evening Post**, September 14, 1946.

buffalo and of the cow country lies on the more irregular lands below the high plains.

The edge of the High Plains, or the edge of the Caprock as it is usually called, extended in a long irregular line across West Texas from north to south. In the South Plains in particular, this great natural escarpment is often the boundary line between cow country and farm land. It was a natural barrier to the grazing of cattle when the first of the little herdsmen came west with their bovine possessions. When the twenty-five cattlemen met at Chimney Creek in 1880, they practically covered the valleys and canyons of the upper Brazos and its tributaries, but not one of them ranched above the Caprock.[5] Some of them, it is true, grazed a narrow fringe of land that extended upon the High Plains but their watering places and winter shelter were down in the valleys beneath. The cattle business had surged against the Caprock and had stopped as a tidal wave might have stopped when it came against a perpendicular cliff.

Here, where my car had just ascended the High Plains and along the outcrop of the Caprock for a long distance north and south of it, was the western edge of the white man's world in 1880. Unfortunately for us in this instance the census taker of that date listed the population by counties instead of by natural subdivisions, but even in spite of this handicap their findings tell a remarkable story about the type of country in which men of the time felt that they could live.

A block of land ninety miles square, the east boundary of which was not more than ten miles west of the place where I had temporarily parked—in other words 5,000,000 acres on the High Plains just west of me—was inhabited by the meager total of thirty-four persons.[6] Nine of the thirty-four lived in Lynn County which was the next west of Garza. Probably all of the nine worked for Colonel C. C. Slaughter who had spread his large ranch far up the Colorado River Valley and also into the broken lands of southeast Lynn County. The other twenty-five of these thirty-four

5. The list of the twenty-five pioneer cattlemen who met at Chimney Creek was published in Chapter 4. The route of their roundup ranged from the foot of the High Plains eastward down the Brazos River Valley. The Stock Manual published in 1881, which included, with few exceptions, the cattlemen of West Texas of that date gave the locations by counties of the various ranches. The writer has carefully checked the entire book and listed these early ranches together by counties. The result shows that ranching extended westward up the Brazos River valley only to the Caprock of the High Plains.

6. The figures, which originated in the U. S. Census of 1880, are taken from the Texas Almanac, 1949-1950, 102-104.

persons lived in Lubbock County, the thirty miles square on the map of West Texas that lay just north of Lynn County. Yellowhouse Canyon, that was spread out like a giant map just east of me here in Garza County, extended northwestward into Lubbock County. Early in 1881 (and probably in 1880) two cow outfits, Kidwell Bros. and Porter Bros., and the Z. T. Williams sheep camp were located up this canyon southeast of present day Lubbock. So far as is known, none of these three stock raising concerns nor any of the remainder of Lubbock County's scant population lived above the Caprock.

But aside from these two instances, the High Plains consisting of this 5,000,000 acre block of flat prairie had not a single human inhabitant that was discovered by the census takers of 1880.

Cold and relentless blizzards swept down across this level plateau in winter and dry thirsty winds licked up the water in its shallow lakes in summer until the cattleman who had accumulated 500 or 1000 cows or more did not wish to gamble his fortune against these uncompromising facts of nature. Timidly and cautiously these frontier herdsmen advanced along the canyons and the draws that led up to the High Plains. They began to learn the few permanent watering places. The draws that led to Yellowhouse Canyon and Blanco Canyon crossed the plains in its shallow water belt. In 1882 Jim Newman moved his 1000 head of cattle to one of these watering places known as the Yellowhouses. It was up Yellowhouse Draw fifty miles northwest of the site of Lubbock in a canyon that offered some shelter to his stock. At about the same time, Sanders Estes and his two brothers began to ranch in Lamb County up Sod House Draw, the north branch of Yellowhouse Creek. A cowman by the name of Lynch from Las Vegas moved his herds to Spring Lake[8] high up the draw that led to Blanco Canyon. Up the same draw, on the part called Running Water Creek, Morrison Brothers began their Cross L ranch at a time not very different from the arrival of these other aggressive cattlemen.

Thus a small beginning was made toward ranching the high flat prairies of the Plains. But cattle raising could not have been successful except around the watering places and along a narrow

7. The U. S. Census of 1880 shows no inhabitants in the Texas counties of Castro, Parmer, Hale, Lamb, Bailey, Hockley, Cochran, Terry or Yoakum.

8. For an account of these early ranchers, see J. Evetts Haley, *The XIT Ranch of Texas.* The Lakeside Press, Chicago, Ill., 1929, p. 49.

margin of land beside the few canyons that cut into this high prairie plateau. There were yet millions of acres too far from permanent water. Nothing short of real pioneering courage could have led a man and his cattle out into those vast stretches of un-watered acres. In many instances the persons who took those great risks were men who sat in swivel chairs a thousand miles away from dry West Texas. One of the early experiments in grazing this doubtful new country lay right beside the road not many miles to the northwest of my momentary parking place.

I resumed my journey along one of the fine smooth highways that are typical of the plains. It was nearly forty miles to Lub-bock across a country almost as level as the ocean. This stretch of road was a continuation of the farming belt that Post had colonized from the old Currycomb Ranch. Hardly had I driven out of Post's former colony, until I had driven into the land once grazed by the cattle of the I O A Ranch (the brand was IOA) which was financed by a group of Iowa capitalists. David Boaz of Davenport, Iowa, came to Fort Worth in 1884 in behalf of these investors and began to buy land and cattle. He purchased and leased not much less than a third of a million acres[9] that amounted roughly to the south half of Lubbock County. Along with this enormous tract he had leased the old Z. T. Williams sheep camp which was seven miles from Yellow-house Canyon southeast of Lubbock. This sheep camp with its rather primitive boxed house became the headquarters of the new ranch. Boaz bought 20,000 head of cattle for a round half million dollars and in so doing put himself and his Iowa consti-tuents pretty deeply into the ranching business. It is doubtful if Boaz knew many of the possible difficulties involved in grazing the flat country above the Caprock. If not, he learned with an additional outlay of hard cash. The great ranch was fenced with posts that had to be hauled from many miles down the canyon. But there was no practicable way of hauling water to supply the thousands of his cattle that must necessarily range on the grass far up on the Caprock—miles away from the springs in Yellow-house Canyon. The problem was met by drilling ten wells and equipping them with windmills— which sounds very simple in view of our present-day understanding of such matters.

But in the middle 1880's the windmill was a comparatively new "gadget," and it took no little courage to depend on it as one

9. For an interesting and excellent account of the story of this ranch see W. C. Holden, **Rollie Burns**, 148-170.

of the absolute essentials of a near million dollar investment. Water wells on the Plains had to be drilled to a greater depth than in the broken country east of the Caprock, and nobody at that time knew that a very plentiful supply of water awaited the drill almost anywhere on the Plains.

Pioneer Texans, wise in the ways of the frontier, would most likely have waited a long time before they spread their ranches over the high Plains in any such manner. Machinery-minded Easterners who perhaps did not fully realize the risks they were taking undoubtedly shortened the time required to extend the cattle business to the millions of West Texas acres that were beyond the reach of living water. The windmill and the barbed wire fence were the right answers to the problem. Without them civilization with the few exceptions already noted would have stopped a long time at the Caprock.

Now as I drove along there was proof of the value of those two inventions as far as the eye could see. The Iowa capitalists sold their cattle after about ten years of unprofitable ranching and finally disposed of their land. Like the Post Colony this area where their herds had once grazed had now become one long succession of well cultivated farms—with barbed wire, windmills, groves of cottonwood trees and farm houses all of which dimmed away into the distant lakes of mirage that are ever present on the High Plains.

By the time I reached Lubbock some twenty-five miles of my road had passed over this land that I O A cows had once grazed. The new city that awaited me with its neatness and civic spirit had turned a cow pasture into the metropolis of the South Plains.

7—Thee Had Better Come Out West

After an overnight stay in Lubbock, I turned east along the highway into the adjoining county. In the edge of Lubbock my highway dipped down into Yellowhouse Canyon for a mile or two before it came back out on the level plains again. It was twenty smooth, level miles to the little town of Lorenza, then a short drive of seven or eight more very level miles off to the north of the highway over dirt roads to the rural village of Estacado. A nice brick school building that belonged to the residents of the local consolidated district was the most important thing in this rural community. It was well that the little village had kept alive the interest in education that had been planted there some seventy years ago. This unique community, however, had its beginning slightly earlier than seventy years ago.

Here was enacted a bit of frontier history that holds a real surprise for those not already acquainted with the story. Almost a decade before there were any such places as Lubbock, Plainview or Amarillo, Paris Cox, a Quaker elder, made his preliminary arrangements to plant a colony of his fellow Quakers here. He purchased 50,000 acres of land and platted a town, when Fort Worth was still the end of the Texas and Pacific Railroad. The town was first named Marietta after his wife, but a few years later the name was changed to Estacado—an identification which it still bears.

Elder Cox began his promotion efforts in 1878,[1] but not until the fall of the next year did he actually have enough followers on hand to constitute the beginning of a colony. The Stubbs family together with the Hayworths, the Sprays, and the Cox family themselves were the little advance group that spent the severe winter of 1879-80 out here on the high bleak plains some twenty air line miles northeast of the site of present-day Lubbock.[2] These families came all the way from Indiana and had perhaps unwittingly become the first farmers to live above the Caprock of the Texas Plains.

Cox had built a house out of blocks of the natural sod that helped his family to withstand the extreme cold, but the other three families who lived in dugouts covered with tents did not fare so well. All of the little colony except the members of the Cox household moved back to Indiana after this one hard winter. But crops were good and others were encouraged to come. Also the idea of a religious center far removed from the influences of more compact eastern communities was very attractive to the Quakers. Doctor William Hunt, a government physician from Indian Territory came in 1880, and in the fall of the year, the Underhill family and the son-in-law George Singer[3] joined the little group of future plainsmen. Estacado grew large enough to become the first county seat of Crosby County; and while it was still a frontier outpost, it had demonstrated the possibility of farming on the High Plains.

But what has the story of this frontier farm community to do with the present ranching outlook of West Texas? As a direct aid to ranching, Estacado amounted to exactly nothing. It was part of the irresistible force that in time was to sweep most of the large ranches off the South Plains. Occasionally early ranchers had found small patches of cotton and other crops very successful but kept the information under cover for fear of an influx of farmers to the Plains. The Quakers, like these ranchers did not wish to be molested in their western solitude, but took no effective measures of suppressing the information that farming above the Caprock was no longer an idle dream.

1. Colonel R. P. Smythe, "The First Settlers and the Organization of Floyd Hale and Lubbock Counties," **West Texas Historical Association Year Book VI**, 19. Paris Cox received his first land patent in this area in 1877, but it was after he discussed the matter with a number of his fellow churchmen that many more land patents were secured early in 1879.

2. John R. Hutto, "Mrs. Elizabeth (Aunt Hank) Smith." **West Texas Historical Association Year Book, XV**, 45.

3. John R. Hutto, op cit.; Mrs. Joe Sherman to J. W. Williams.

These gentle folk from Indiana did not build their homes along the canyons as others had done where living water might have become the center of their tiny colony. On the contrary they had struck out boldly on the High Plains, dug ninety feet into the earth for water,[4] and had set out to plow the land even before the first cattlemen of the South Plains were willing to venture above the Caprock. By 1882, when Sanders Estes and Jim Newman had moved their cattle to the upper branches of the Yellowhouse, the Quakers were so well established that they were already beginning to provide school facilities for their children.

These mild spoken Easterners had proved, well in advance of the big ranches that the high plains country was, in a measure at least, adaptable to farming. But in spite of that demonstration, in due time the ranches came and covered most of the grassland above the Caprock. By 1886, the giant X I T Ranch[5] had spread itself across ten[6] West Texas counties. By the aid of barbed wire and windmills, the financial backers of this mammoth undertaking conducted the world's greatest fenced ranch—much of it above the Caprock where cattlemen had been afraid to venture. The windmill and barbed wire—and enough capital to carry the business to completion were essential factors with the X I T. The project was financed with foreign and eastern capital. With a less plentiful source of money, Nunn Brothers who had grown to giant size in the cattle business since 1866 found the unusual expenses of ranching above the Caprock too great when they moved into Terry County about the time that the X I T began. Nunn brothers with their famous N U N brand were forced to retire from the ranching business.[7] However, the added expense of supplying water on the High Plains and other costs and drawbacks that were peculiar to that section did not prevent the grazing industry from covering most of this West Texas plateau. Then, possibly because of increased land values, some of the big ranches

    4. Colonel R. P. Smythe, op. cit., 20.
    5. The first herd of cattle actually came in 1885. J. Evetts Haley. **The XIT Ranch of Texas,** 81.
    6. Recently the accuracy of this statement has been called in question. The claim has been advanced that the XIT Ranch extended into only nine Texas counties, leaving Cochran County out of the original list of ten. However, the Texas General Land Office map of Cochran County for December 1913 shows that leagues No. 700 and 701 of the State Capitol Land are partly in Cochran County. Perhaps 4000 to 5000 acres of the land surveyed for the giant ranch lay in Conchran County. The fact is confirmed by the maps of present day commercial map makers.
    7. Gus L. Ford, **Texas Cattle Brands,** 77. The NUN cattle were sold in 1889 and the purchasers had disposed of the herds by 1891. The late J. T. Bond of Jayton, Texas, who supplied the Ford publication with this information referred to the **NUN** cattle venture as a financial failure in an interview with the present writer.

sold their lands out in large blocks to independent cattlemen. They were not able to cut these great pastures into farms until enough farmers had come to the plains to make the land marketable.

Herein lies the significance of the little village of Estacado. Farming began to spread from the Quaker colony as a beginning point. Especially did it show early gains a few miles north of Estacado, in Hale and Floyd Counties. By 1890, these two counties and Crosby, which contained the Quaker Colony, had a total population of 1596,[8] a small beginning to be sure, but even this was at least five times the population that would have been required to carry on ranching in the three counties.

To make clear the full import of this modest total of 1596 persons who had moved into Hale, Floyd and Crosby counties by 1890, let us draw a line on the map across the High Plains west to New Mexico, thirty-five miles south of Amarillo, and another line west to New Mexico beginning twenty miles northwest of the town of Big Spring. Now draw another line from near Big Spring northward along the irregular course of the Caprock and draw still another line southward along the east boundary of New Mexico. These four lines would enclose roughly a great rectangle 150 miles long north and south, and from eighty to 120 miles wide, east and west. Sixteen counties that cover 15,000 square miles are in this big enclosure. They include the heart of the South Plains, and for our purpose comprise an area where population figures for a long time were not materially influenced by the coming of railroads.

Now take your knife and cut out this part of the Texas map and paste it down on a flat block of wood. Take care to cut the map along county lines near the edge of the Caprock instead of following the Caprock exactly, since census figures which we must use in a moment are given by counties instead of by natural subdivisions. Now put your thumb down on this cut-out map on the corner of Crosby County northeast of Lubbock, allowing your thumb to extend over parts of Hale and Floyd counties far enough to cover both Floydada and Plainview. It was that little spot on the map under your thumb where most of the 1596 people lived who were shown as residents of Hale, Floyd, and Crosby counties in the census of 1890.

Before taking up your thumb, look for a moment at the rest of that cut-out map. All of the other thirteen counties, covering

---

8. The **Texas Almanac**, 1949-1950, 98-104. Other population figures in this chapter are taken from the same source.

an area of at least twenty times as large as the spot under your thumb, had the small total of only 299 inhabitants when the census takers of 1890 counted noses. Plainly on the map of the High Plains this small spot that spread northward from the little Quaker village of Estacado was the modest beginning of a farm belt. It contained the germ of destruction (if such it may be called) that in due time was to eat the big ranches that covered the rest of the plains into ten thousand little pieces.

Ten years later in 1900, when the census takers checked up again, that small spot on the map had grown some—it was a little too large to cover with your thumb. The miniature farm belt had spread to Swisher County, which was just north of the town of Plainview. The four counties, Crosby, Floyd, Hale, and Swisher now had 5700 inhabitants while the other twelve counties on your cut-out map an average of just one person to each ten square miles of land. Without any argument, ranching was still supreme on the South Plains.

But the day of sweeping changes was at hand. Railroads were about to extend cheaper transportation to most of this land above the Caprock. Thousands of covered wagons loaded with farm families and their scant belongings were about to head west toward the high plateau where a little handful of Quakers had found it possible to live by cultivating the soil—not that they had heard so much about the Quakers as about the few thousand Texans who had repeated the demonstration begun by the Quakers.

By the time the 1910 census was taken, the little farm spot had grown until it showed signs of covering at least eleven of the sixteen counties on your cut-out map. Five of these sixteen counties that lay between Lubbock and New Mexico still had rather small populations, but none of the other eleven had as few as 1000 inhabitants. The part of the High Plains represented by your little map now had 33,000 residents and ten years later it had a round 60,000. The nester—the irresistible force that was due to tear the big ranches to parcels—had arrived on the High Plains.

So far we have told in sweeping statistics about this movement to subdivide large ranches across this span of sixteen counties— but the story may be broken down into many specific instances. There is room here for only a little of the detail. The great X I T Ranch, with its 3,000,000 acres spread all the way down the west edge of the Texas map from the corner of the

Panhandle to the middle of Hockley County west of Lubbock, was due to share in this period of ranch disintegration. Something like 1,000,000 acres of the ranch was within the sixteen counties under immediate consideration here. The south end of this million acres, which was in fact the south end of the ranch as a whole, was sold to Major Littlefield and associates in 1901. The sale, at a price of $2.00 per acre, included 235,000 acres,[9] nearly all in Lamb and Hockley Counties. This land centered around the famous watering place up the south branch of Yellowhouse Canyon known as the Yellow Houses where Jim Newman had begun ranching in 1882.

Jim had purchased a little adobe hut at these springs from the Causey boys,[10] the last of the plains buffalo hunters, and had made it his headquarters until 1885. Newman, in accordance with the open range custom of his day, had not bothered to become owner of the land and, of course, was forced to move on when the Capital Syndicate purchased it as a part of their famous X I T Ranch.

The new occupants of the land adopted the spring at the Yellow Houses as one of their headquarters and continued to use it until the sale was made to Major Littlefield. The Littlefield interests erected a group of buildings at the spring and painted them yellow in keeping with the color of the natural rock formation which formed a part of the canyon wall—said to have resembled yellow houses at a distance. When Littlefield took it over, the spring was beginning to fail, so a well was sunk nearby, and a windmill with a tower 128 feet high was put into operation.[11] The tower, tall as it was, reached only about 30 feet above the high walls of the canyon and, of course, reached up into much swifter currents of air than a mill of traditional height.

Littlefield and his partners prospered at the new ranching venture. They produced $1,000,000 worth of cow flesh during their first ten years,[12] but even brighter days were ahead. Late in 1910, the Santa Fe Railroad built across the corner of their ranch and land values began to spiral upward. The town of Littlefield, on part of their land, was launched thirty-six miles northwest of Lubbock and profits that were too tempting to resist faced the

9. J. Evetts Haley, The XIT Ranch of Texas, 221. The exact amount of Littlefields purchase was 235,858½ acres.
10. Ibid, 49.
11. Orville R. Watkins, "Hockley County from Cattle Ranches to Farms," West Texas Historical Association Year Book, XVII, 47-48.
12. Ibid, 48.

ranch owners.   Thousands of acres were sold at prices that ranged from fifteen to thirty-five dollars per acre.[13]

By 1923, the ranch was sold out—all but a remnant of 23,000 acres that snugly fitted into the canyon at the Yellow Houses.[14] In a few years the yellow buildings burned and the 128 foot windmill tower blew down, but both have since been replaced by better equipment.   A son of Phelps White, who was Major Littlefield's nephew and partner, now runs high grade Hereford cattle on the ranch.   His fine brick home near the same place where Jim Newman lived in the old adobe house is done in yellow, still in keeping with the tradition of the canyon—a tradition older than written history.   An ancient watering place of the Staked Plains and a nice modern ranch have been preserved, but most of the near-quarter million acre spread of the old cow country that was taken over by Major Littlefield almost a half century ago has given away to the nester and the plow.

Along the east edge of this extensive tract of land and stuck to it like a postage stamp, was assembled another of the large South Plains ranches.   In 1885, about the time when the X I T management was beginning to assume possession of its far flung landed empire, Dudley Snyder bought from Lewis O. Nelson a strip of land eight miles wide by twenty-five miles long along the east border of this southern part of the X I T holdings. Dudley's brother John soon became a partner in the transaction, and not long afterward, the partnership purchased another seventy-one square miles that joined them on the south. The brothers had paid $1.65 per acre for the first tract and $2.00 for the second and were not averse to taking a profit.   Colonel Ellwood, the barbed wire king from DeKalb, Illinois, came to the Plains country in 1889 and offered the brothers $2.50 an acre for their first tract of land. It meant nearly $110,000 profit to the Snyders and the deal was made.   Thirteen years later Elwood purchased the second tract for $3.00 an acre, paying altogether more than $150,000 in profits to the Snyder brothers.   But Ellwood was still land hungry, and in 1904 he acquired four leagues of school land at about $3.00 per acre; in 1906 four additional leagues at nearly twice the price; and sometime later another four leagues at about three times his first purchase price.   Altogether the barbed wire magnate had spent nearly three quarters of a million dollars for land, but he

13. J. Evetts Haley, **George W. Littlefield Texan**, 199.
14. Orville R. Watkins, **op cit**, 48.

had built up a ranch of some 225,000 acres.[15] His immense estate was 54 miles long, extending some distance past the south end of the former X I T Ranch. Colonel Ellwood had purchased a herd of cattle, bearing the spade brand from J. F. Evans of Donley County, at the beginning of this ranching venture and had kept the brand as his own. It is from that beginning that the great estate became known as the Spade Ranch.[16]

The rise in land values, that had made the dissolution of the Littlefield Ranch too tempting to refuse, acted similarly on this neighboring Spade Ranch. Colonel Ellwood sold the north end of his long ribbon shaped holdings to farmers and moved the headquarters of his shrinking empire toward the south. As was true of the Littlefield Ranch, this was another case of invasion by the plow.

In 1906 near the center of Hockley County south of Littlefield's ranch and west of the Spade, the same C. W. Post who was soon to found Post City bought the Oxsheer Ranch. The real estate boom that struck the north portions of the Littlefield and Spade properties in 1910 did not immediately spread to central Hockley County; but Post, who dreamed of colonies of farmers instead of acres of cows, made ready for the boom when it came. He platted a town on his ranch and named it Hockley City.[17] The name failed to survive, but the town plat lived, and on it grew the county seat and principal town of its immediate area—fifty percent larger than Post's own Post City. Here, thirty miles west of Lubbock in the heart of the South Plains on land perhaps as level as any on this planet, a town grew up just as Post had had his dream on paper more than a decade earlier. Quite appropriately the new town was called Levelland. Around it Post's holdings were put on the market in 1925; then, only a year later, the sale of the Slaughter Ranch to the west began.[18] Thus the great grassland empires continued to fall.

To the north of Littlefield, the X I T interests sold 185,000 acres to William E. Halsell[19] the same year that the sale was made to Littlefield. When Halsell stocked his new ranch, he revived the old circle brand that he had run in Clay County twenty years

15. Ibid, 49.

16. Ibid. J. F. Evans was ranching on White Fish and Saddlers creeks in Donley County as early as 1881. His address at that time was given both as Clarendon and Sherman, Texas. See George B. Loving, The Stock Manual, 30.

17. Orville R. Watkins, op. cit, 51.

18. Ibid.

19. J. Evetts Haley, The XIT Ranch of Texas, 221.

earlier except that he flattened the circle[20] at the top and bottom until it looked almost like a link out of a chain. The cowboys called it the "Mashed O" and under that name and brand the ranch still survives. The Halsell headquarters are at Spring Lake in northwest Lamb County where the X I T owners had long maintained a camp. Like most of the other large ranch properties, thousands of acres of the Mashed O grasslands have surrendered to the plow. Something near half of the original purchase has become farm land. Lamb County, most of which Littlefield and Halsell long held within the loops of their great lariats, now stands second from the top of the South Plains counties in the amount of cultivated acres.

Some seven or eight miles north of the Mashed O Spring Lake headquarters is the southeast corner stake of Parmer County. This High Plains county, pushed as it is snugly against the east line of New Mexico, is a square block of land that runs thirty miles northward and westward from the above mentioned corner. It is the northwest 550,000 acres on your cut-out map, and about 98% of it once was a part of the X I T Ranch.[21] Now, this former piece of pasture land has been subdivided, and 385,000 acres of its sod[22] have been turned and given over to the planting of crops.

All in all, the sixteen counties on your cut-out map now have above five million acres in cultivation.[23] To realize the full import of these figures, compare this block of High Plains land with 15,000 square miles of the great farm belt that includes Dallas, Fort Worth, and Waco. This latter block including fifteen counties, roughly the same area as the above sixteen on the High Plains, has just less than four and a half million acres in cultivation. Even including the still thinly populated counties at the western margin this part of the High Plains, this former ranch land above the Caprock now has 10% more acreage under the plow than has the same area in the heart of this most highly developed farm section in Texas!

The 15,000 square miles of the High Plains that had 60,000 inhabitants in 1920 grew to 173,000 in 1930 and to 200,000 in 1940.[24] With this tide of immigration still sweeping away the great

20. Gus L. Ford, **Texas Cattle Brands**, 139. The Halsells had run the mashed O brand in Oklahoma as early as 1889.

21. See map of Parmer County, General Land Office, Austin, Texas, May, 1926.

22. The Texas Almanac, 1949-1950, 575.

23. The Texas Almanac, 1954-1955, 214-216.

24. **Ibid**, 98-104.

ranches, you perhaps may ask again, "What are the future prospects as to beefsteak as a part of the American diet?" But again let us insist that much of the evidence is farther down the road. A conclusion now would still be premature.

Just as is the case completely across this part of the map, here at Estacado where Paris Cox and his mild mannered Quaker adherents broke the first sod of the South Plains, the evidence also seems all against the probability that beefsteak will endure as a plentiful commodity. Here, just as in many other parts of the plains, there are farms, farms, farms—as far as the eye can see— good rich black dirt that, with enough water, can produce almost like the valley of the Nile! But as one stretches his vision to the very limit, not a ranch is in the whole scope of view. But do not let the one-sided argument overwhelm you until the old car has brought us down the road far enough to find all the facts.

I had parked the car in front of the little store at Estacado and had begun to look around. Should I sit here awhile and observe? Perhaps a pair of Quaker maidens might appear—in traditional headdress such as one finds in story books. One of the maidens might invite the other into the store with the friendly suggestion, "Thee had better have a cold drink, thy thirst is evident." Then, eavesdropper fashion, I might slip in and hang on one of the counters of the little store and listen to the two young ladies converse in language that perhaps would suggest that they had just stepped from the pages of a King James Bible. But I was disappointed. In a few minutes a rather smartly attired young lady did enter the store and I followed, halfway hoping that she was a Quaker maiden who had neglected to dress in costume.

Once inside the store I engaged her in conversation. She was not a Quaker—she was a teacher at the local school. Yes, she knew something of the story of the old Quaker Colony: it had broken up about 1890 and had scattered in many directions. One of the Quakers, a son of Dr. Hunt, had become president of McMurry College at Abilene. Others had played more or less prominent roles in the history of the West. One of the Underhill girls was still living. She had married and had become a true to type Westerner. She had become Mrs. Joe Sherman of Seminole, a town even farther out on the High Plains than Estacado. Then came the most startling statement of all—not a Quaker was left anywhere near this little pioneer village.

Here had been enacted one of the strange bits of human drama. Paris Cox and his brother Quakers had come far out to this place on the lonely western plains hoping to bring up their families unspotted from the sinful world. But instead of remaining aloof from the world as they had thought, the world gradually moved to them, and they mingled with it, and became an integral part of it. In some instances they took the lead toward a better state of living, and as a group they proved by courageous example that human beings could earn their living on these high bleak prairies, and that the land would produce something beside beef steak.

8—The End of the World in 1877

I drove back down the road to intersect my highway at the town of Lorenza which, it will be remembered, is twenty miles east of Lubbock. The pavement continued eastward here still over the monotonously level prairies. On both sides of the road was one long succession of farms—some of them irrigated and consequently covered with crops of darker green than the dry land farms beside them. It was seventeen easy miles to Crosbyton. This place is the end of a roadroad that runs eastward from Lubbock. Another railroad that extends northwestward from Stamford ends at Spur not more than thirty air line miles southeast of here. The two roads are pointed at each other almost like rifle barrels. One might look at a map of the area for a long time and wonder why the two railroads were not connected. What the map probably fails to reveal is that a few miles east and southeast of Crosbyton is the edge of the Caprock of the Plains, a near perpendicular cliff several hundred feet high. To make an ascent up a cliff is no small problem for the engineers who plan railroads. And unless the possibilities for profit are exceptionally strong, investors may decide against such an outlay of capital. It was the same barrier that, as we have seen, had marked a line across the Texas prairies in more ways than one. It was here in Crosbyton that I had on a previous visit met George Smith who had invited me to his home out in Blanco Canyon north of town.

Stories of the Smiths and their canyon home had long fascinated me, but I had never paid them a visit. George's father and mother were two of the most interesting pioneers known to the early West and their old rock house, once the western outpost of civilization and now a historic landmark, was already beginning to gather legends. It was seventy-seven years ago that Hank Smith and his remarkable wife, Elizabeth, had pointed the tongue of their covered wagon up Blanco Canyon toward this place as their future home.

Hank had come to America from Germany when he was only twelve years old. He had lived three years in Ohio, but the call of destiny beckoned him to the wild Southwest. In the strange free life of the frontier he was schooled to become one of the important citizens along the fringe of the Texas settlements. Both as cowboy and soldier he had evidently done his work well. It was with Sibley in New Mexico that he had worn the gray uniform during the Civil War.

Elizabeth Boyle, his wife, quaint in her old world manner but intelligent, and adaptable to new surroundings, had come fresh from Scotland to the West Texas frontier. Elizabeth and Hank met at a dance at Fort Griffin in 1873. After a short courtship they were married and settled down in Fort Griffin, perhaps planning for the day when the Texas Plains were to be cleared of wild Indians. It was just four years after they met at the dance that the young couple drove their covered wagon up Blanco Canyon to establish the first home on the South Plains.[1] Their first two children were in the wagon. The older of the two was a boy named George.

Their new home was ready before they arrived. It was built by a young Easterner by the name of Tasker whose bragging manner had made him the laughing stock of Fort Griffin. Tasker had become heavily indebted to Hank Smith, and Hank had to take this property to satisfy the debt. But whatever may be said of Tasker, one must admit that his house was well constructed. It was two stories high and of solid rock masonry.

When the Smiths moved into their new home, Fort Griffin 160 miles east was their nearest trading point and post office. Rath City, a temporary trading post for buffalo hunters, was only 75 miles down the road, but the buffalo hunt was about over and the place soon ceased to function. The Hank Smith rock house was the last human habitation west on the way from Fort Griffin

---

1. John R. Hutto, "Mrs. Elizabeth (Aunt Hank) Smith," **West Texas Historical Association Year Book**, XV, 40-47. This account of the pioneers Mr. and Mrs. H. C. Smith is based upon the Hutto article and supplemented by my own personal interview with George Smith.

to New Mexico. It is because of this fact that the old Smith home deserves mention in this story of cattle and ranching. It was the western anchor point to which ranching, farming, or the other doings of civilized white men were fastened.

It was from the Hank Smith rock house in Blanco Canyon that Paris Cox and his Quakers set out upon the High Plains to establish Estacado. It was Hank Smith who dug the first water well for these future plainsmen; it was he who turned their first thirty acres of sod and made their demonstration of farming on the plains a possibility.

Hank was not, as such, an important cattleman. His herds never reached more than about 600 head, although he ran one of the most interesting brands of his area. His cattle wore the Cross B brand,[2] suggested perhaps because he was the first settler in Crosby County.

In 1879, the Hank Smith home became the Mount Blanco Post Office. As the home had been the last of such places west, so the post office became the end of the mail routes from the East. "Aunt Hank," as Elizabeth became known to the cowboys, was the first and only postmistress.[3] Her official duties in that capacity did not cease until 1916—four years after the Santa Fe Railroad had built through Lubbock.

Hoping to learn more of this unusual Smith family, I drove northward out of Crosbyton toward Blanco Canyon. The country to the east of my road had once been the property of the Two Buckle Ranch, more accurately called the Kentucky Cattle Raising Company. It was on that range that our friend Red Mule Barkley of Clairmont, had worked and in some way had acquired his colorful title more than fifty years ago. The old ranch had long since gone out of business, but a part of it had more recently become one of the excellent ranching properties of J. S. Bridwell of Wichita Falls, Texas. Bridwell owns some ten miles or more of Blanco Canyon extending from the highway east of Crosbyton almost to the old Hank Smith home far to the north.

Nearly ten miles north of Crosbyton I drove down off the plains into Blanco Canyon and passed along the west side of Bridwell's ranch. It was not far to the old rock house into which Hank and his family had moved just seven decades ago. Both Hank and Elizabeth had died a number of years ago and their descendants now had charge of the property. The home was now

2. Gus L. Ford, **Texas Cattle Brands**, 91.
3. John R. Hutto, op. cit, 45.

occupied by a grandson, Bob Smith, son of the late Blanco Bob Smith who was the first child born in the old house. The family of Blanco Bob now lives in Lubbock, but his son and daughter-in-law have charge of the old home. It was young Bob's wife who very courteously showed me the home. She pointed out the south window of the old hallway where "Aunt Hank" had so often handed out the mail. The old house made a very impressive appearance. Its walls were more than twenty inches thick with not a crack in evidence after standing seventy-five years. The old rock fence that enclosed the house and yard was still standing. Not far behind the east chimney were two large pecan trees which Hank had planted. The trees had grown vigorously, even though they were 150 miles west of the nearest pecan trees that had grown naturally. These two trees matching the height of the old two-story house gave the place a dignity befitting its years.

I drove two or three miles east, passing between the Smith property and Bridwell's Ranch, until I came out on the plains again. It was not far to George Smith's home. Most of his land was down in the canyon, but his house was out above the Caprock. George was down the canyon looking after some sheep. I awaited his return to the house, expecting to interview him and go on my way; but both he and Mrs. Smith insisted that I stay for lunch.

I enjoyed their hospitality for several hours. Mrs. Smith showed me many newspaper clippings that she had treasured through the years and called attention to a very lifelike picture of Hank and Elizabeth as they sat before the old fire place. George pointed out a heavy "square barreled" old gun that stood in the corner of the room. It was Hank's old buffalo gun.

The hours passed too fast, and, before I knew it, it was time to leave. George guided me over the canyon that had been his home since he had come up from Fort Griffin in the covered wagon with his mother and father some seventy-five years ago. Coming down into this natural recess in the plains a half mile south of his house, was the old road where the mail had passed. A mile southwest of the old Hank Smith house, the Mackenzie Trail climbed out of the Canyon. As we drove along, he told me the story of the canyon through the years, almost as fresh as if the stirring incidents of his narrative had occurred yesterday.

I finally and reluctantly broke away from his interesting guidance and went back to Crosbyton over the same road and over the same level farmland that I had followed to the Hank Smith home earlier in the day. Crosbyton had, after a long time, become the county seat of Crosby County, but it was not the first. It will be remembered that the Quakers organized the county with Estacado as its county seat. Paris Cox, their leader, was the first county clerk.[4] Then the population began to spread, and old Emma which was much more centrally located became the county capital. Next came the thirty-seven mile railroad from Lubbock, and Crosbyton at the end of this short road assumed the official leadership of the county. This shifting of the court house was not just a story peculiar to Crosby County. New counties all over the West had their day of uncertainty in this regard as the status of ranching, farming, and new railroad building changed the map and at length made it into a permanent pattern.

Three miles east of Crosbyton I again drove down into Blanco Canyon. It is the same Blanco Canyon up which Hank and Elizabeth Smith had driven the old covered wagon from Fort Griffin when they were en route to occupy the rock house ten miles north of my road. It was the same canyon up which Mackenzie had pushed his first campaign against wild Indians in 1871. It was the same canyon into which rushed the many little cattlemen who approached the plains in 1879. It was up this canyon that civilized white men had made their first approach to this part of the great South Plains. Here, in my very road and spread over quite an area to the north and south of it was the old Two Buckle Ranch that lasted from the middle eighties to near the turn of the century. Their old headquarters had been somewhere down in the canyon ahead of me. Now to my left and up the canyon all the way to the old rock house is the fine ranching property of J. S. Bridwell mentioned above.

Blanco Canyon at this place is some four miles in width and perhaps three hundred feet deep. Heavily laden trucks glide down into it with ease, but some of them struggle no little in making the ascent to the other rim. The creek at the bottom of the canyon, variously called White River, Catfish Creek or Running Water Creek, is really the principal branch of the Brazos River, but history has chosen not to so name it. This stream unites with what is known as the Salt Fork of the Brazos some thirty miles southeast of here in West Kent County. The so-called Salt Fork (or principal river) is only about thirty miles long above that junction of the two streams, but this branch that flows through Blanco Canyon is more than two hundred miles long and has its

4. Ibid, 46.

source on the Plains of New Mexico northwest of Clovis. But even though the discrepancy between names and geography is large, the names probably will remain as they are.

On the east side of Blanco Canyon the highway follows the level plains again. Ten miles of the smooth highway, again through a farming belt, brought me to the east edge of the Caprock. It was seven or eight miles farther to the little county seat town of Dickens. Here I turned south ten miles to Spur and spent the night at the Spur Inn.

9—Sixty Years from Scotland

This hotel is more the type one would expect to find at a health or pleasure resort than out in the old West Texas cattle country. Its large lobby blends into its still larger porches to make the entertainment area of the quaint old place seem endless. The town of Spur with a little more than 2000 population is in the heart of a farming and ranching section with a most interesting background.

Even before the big buffalo hunt swept across the foothills of the Plains, the Houston and Great Northern Railway Company sent their surveyors into this section to measure off something like a half million acres of public domain. To them, the land came as a premium from the State of Texas for miles of railroad which they had constructed in the settled areas of the State. Quite a little of the surveying was completed as early as 1873,[1] but there was no market for these lands that were still covered with wild buffalo and marauding Indians. The immense property was transferred to the New York and Texas Land Company and re-surveyed in 1882.[2] We have seen how the obstacles to cattle raising were removed before 1880, and we have noted also that

---

1. Dickens County sketch file 9 in the Texas General Land Office shows block 1 and 2 of these H. & G. N. lands. The sketch was filed August 13, 1873.
2. The Scurry County surveying records contain an old volume of surveys done in Kent County. Many of these are the H. & G. N. sections of land surveyed in 1882.

the frontier cattlemen were not at all timid about moving in and helping themselves to the grass regardless of where the surveyors may have left their stakes.

Many of the cattlemen who met at Chimney Creek in 1880 were grazing their herds on Houston and Great Northern grass. Chimney Creek itself was on their land, but absentee ownership meant nothing to the man on the open range. There was no attempt on the part of the owner to take over the land, and it was custom all over the range for cattlemen to respect the rights of the man who first moved his herds into any particular area. But this golden age and its free use of grass was of short duration in the Spur country.

In 1883, A. M. Britton, S. W. Lomax and others formed the Espuela Cattle Company and purchased at about $2.00 per acre this great range[3] that had now become the property of the New York and Texas Land Company. Actually the purchase covered only about a quarter of a million acres, since only the alternate sections had belonged to the railway company. The other half, checkerboarded with the railway lands, belonged to the Texas school fund. But regardless of such complications, the new company took over this great block of land and purchased cattle to cover their seemingly endless domain.

George Gambel, Bud Campbell, and several more[4] of the twenty-five cattlemen who had formed the Chimney Creek organization transferred their cattle to this new company. There were numerous other transfers, but probably the first of all these purchases and certainly the most significant was the deal to acquire the herd that had belonged to Jim Hall of Richland Springs, Texas.

Hall had for a long time branded cattle on his range near Richland Springs with the Spur brand.[5] He traded in cattle enough to have been possessed of other brands, but the spur seems to have taken precedence whenever practicable. By 1877, Hall began to shop around for new range. He and his brothers moved up into northeast New Mexico, but within a year he sold out his interest there and came back to Texas.[6] He became interested in the vast stretch of open range in West Texas that had been made available by the slaughter of the buffalo.

3. William Curry Holden, **The Spur Ranch.** The Christopher Publishing House, Boston, Mass., 1934, pp. 15-17.
4. W. J. Elliott, The Spurs. The Texas Spur, Spur, Texas, 1939, p. 33.
5. **Ibid,** 25.
6. William Curry Holden, **The Spur Ranch,** 13-14.

Fifteen hundred head were purchased by him down in south Texas and trailed by his men to Motley County where he, after some delay, located on the head of the north Pease River northwest of the present-day Matador.[7] In addition to these cattle that came from the coast of Texas, a small herd was purchased from his brothers in New Mexico and driven across the Plains to his Pease River Range. Hall maintained his home at Richland Springs, but for four peaceful years his hired hands made cowboy music and his cattle added pounds of cowflesh out here at the foot of the Plains by the Quitaque Peaks. Undoubtedly his cowboys burned the Spur brand into the hides of his recently purchased cattle as well as of his calf crop, for the old faded brand book of 1881 lists two of his advertisements, both of which feature none other than the Spur brand.[8]

This scene of peaceful growth enacted out here by the rim of the Plains was disturbed soon enough, for the newly formed Matador Cattle Company had bought the very ground under his feet. In 1882, he moved his increased herds, then amounting to perhaps six thousand head, to Red Mud Creek some ten miles southwest of the present town of Spur. My good friend, the late Jake Raines, of Sweetwater, Texas, was one of Hall's cowboys who helped with this now famous drive from the Pease River to the grassclad valley of Red Mud Creek.[9] Most likely, to make his new range secure, Hall purchased the Double Block and N Bar brands of cattle that were already nipping the grass along the banks of the Red Mud.[10] The range went with the cattle and the owners of these two brands of cows, J. W. Hall (not related to Jim) and his partner Montgomery, seemed willing enough to make the deal. Jim Hall's cowboys got out their branding irons and burned the Spur brand into the hides of the newly acquired cattle, and thus the symbol of the Spur was more firmly planted in this great empire of grass that had been surveyed for the Houston and Great Northern Railway. It soon became the name of the very town in which I was spending the night.

Shortly, Jim Hall sold his combined herds to Stephens and Harris, and they in turn sold them to the Espuela Cattle Company[11] that was about to purchase the immense block of railroad land that stretched for miles in all directions from this present-

7. Ibid, 13.
8. George B. Loving, The Stock Manual, 138, 246.
9. Personal interview with Jake Raines.
10. William Curry Hollen, The Spur Ranch, 15.
11. Ibid

day town of Spur. This first ranch deal was completed in 1883, but there were yet other milestones to be passed. In 1884 the combined holdings were transferred to the Espuela Land and Cattle Company of Fort Worth, but after that the transaction reached far beyond the limits of Texas. The entire ranch was again sold in 1885—this time to the Espuela Land and Cattle Company, Limited, of London.[12] The principal investors this time were a group of Scotchmen from Aberdeen, Scotland, and nobody who knows both cowboys and Scotchmen can fail to smile a bit at the new combination.

Certainly there were mirth provoking occurrences, but West Texas has acquired some excellent citizens as a result of that strange new partnership. One of those fine citizens who came after the London Company had been operating the ranch for two or three years was the late W. J. Elliot who in much of his later years, lived on Red Mud Creek ten miles southwest of Spur. Elliot never entirely lost his Scotch brogue, but one must also observe that under his fine Scotch skin had grown a good Texas cowboy to match the high class, intelligent gentleman that was under that same skin when he first came to the West.

These Scotchmen from 5,000 miles over the sea, aided by some London financeers, built up a great ranch and continued to operate it for twenty-two years. They sold their land, cattle, horses and equipment to the Swensons and associates[13] of New York City in 1907. The market for West Texas land was very noticeably on the up grade and the Swenson interests made their purchase at the right time to profit by a rising market. They saw to it that a railroad was built out into the heart of the ranch and they laid out the town of Spur at the end of the new road. The tide of west bound immigration that was about to sweep over the High Plains also came to the Spur Ranch. Like C. W. Post the Swensons also had their dream of a colonial empire out at the foot of the Plains. The dream came true and the little town of Spur was born, and with it, hundreds of farmers began to cut down the mesquites and to turn the sod right where the successors to Jim Hall's spur-branded cattle had cropped the grass for a quarter of a century. The new farm belt stretched out the full twenty-four miles southeast to Jayton where I had spent the night forty-eight hours earlier. The old Scotchman, with the Two Circle Bar brand and the unclipped beard, over in what is now the Jayton country had

12. Ibid, 18
13. Personal interview with A. J. Swenson of Stamford, Texas.

joined the Spur Ranch on the east side back in the days when there was little thought of this nester invasion. But in due time the nester came and the old cattle kingdoms began to crumble.

The Swensons, however, even until this day, operate a very large remnant of the old Spur Ranch. The old west pasture at the mouth of Blanco Canyon and quite a little other range along with it is still a part of their extensive holdings. Their cows still graze historic little Chimney Creek where the open range herdsmen first met in 1880, although the valley of Duck Creek, principal stream of the original east pasture, has now for more than forty years produced cotton and other row crops.

Ranching was not a new venture for the Swensons when they came to the Spur country. For a quarter of a century they had grazed an extensive spread of acreage that was located in four or five West Texas counties. Soon after 1880, they began to operate the Ericsdale Ranch of 110 square miles that lay between the city of Stamford and the Clear Fork of the Brazos. Shortly they started the Flat Top Ranch of somewhat greater area that occupies the smooth crest of Flat Top Mountain northwest of Stamford. About the same time, the Swenson herds were spread over another large ranch that covers most of the northwest quarter of Throckmorton County.[14] At the time this is written, they have run the well known S M S brand on these last two ranches for full seven decades. About a year or two after the turn of the century, the Scab Eight Ranch on the Tongue River between Matador and Paducah became Swenson property[15] and adopted the Swenson brand. And now, as we have seen, the giant Spur Ranch was purchased in 1907.

The Swensons not only set about to market a part of this newly acquired land, but they began to change over from the old brand of the "Spurs" to their own S M S brand. Their last herd of 20,000 Spur branded cattle was sold to W. J. Lewis of Dallas[16] in 1910. Now nearly eighty years after Jim Hall started this famous brand, Lewis still runs it on several thousand fine Hereford cattle on his ranch in Hall County not so many miles from the Quitaque Peaks where the brand first came to this part of the West.

I slept rather late at the Spur Inn and didn't bestir myself

---

14. The A. J. Swenson interview.

15. The record books in the Swenson offices at Stamford, Texas show the dates of these land transfers.

16. Gus L. Ford, **Texas Cattle Brands**, 130.

with much speed after I was once aroused. It was my plan to call on W. J. Elliot who had come to America unofficially as a trusted representative of some of the Scotch investors in 1888. Elliot had made many friends among the old cowboys whose ranks are thinning now all too fast. On the range he was known as "Scotch Bill," but the old cowhands who still remember the name are very few indeed. It was my pleasant privilege to carry greetings between this fine old Scotchman and Jake Raines a short time before Jake's death. Elliot wrote a book ("The Spurs") which should be in the library of those who love a good true story of the old West. I was the messenger who took a complimentary copy of this book from Bill to Jake—and the pleasure was not altogether confined to these two old cowboys!

I drove the ten miles from Spur to Bill Elliot's ranch, about the middle of the morning. At the time this journey was made he owned a small tract of land in the valley of Red Mud Creek near where Jim Hall first drove his spur-branded cattle. The residence was at least a quarter of a mile through mesquite timber from the front gate, and here I found a unique combination of plow tools and culture. The three daughters are graduates of leading institutions of learning, but along with it they know how to shift gears on a tractor. They also know the fine points of scholarly research, and Bill himself had assembled archaeological materials that will some day make a fine acquisition for some University.

After he showed me his interesting collection, and we talked of the old Spur country, I drove back over my route to Spur and then back to Dickens.

Suppose we park the car here for a while and review some of the colorful hair-raising tales that linger in this part of the big ranch country. In due time we shall return to this journey through the great ranges of the Southwest.

10—Cow Country Gossip

A lone horseman was kicking up the dust at a fast clip, riding toward the old Bar X headquarters. Neither he nor his dust were easy to see because the sky was in partial darkness even though it was well toward the middle of the day. Dick Jones, the sixteen-year old cowboy astride the fast moving horse had never seen an eclipse of the sun before—and to put it mildly, he was frightened. His body slipped to the ground feet first beside the corral fence where he stopped the panting horse at the end of his hurried ride. Pete Smith, the wagon boss who was momentarily at headquarters saw the kid ride in.

"Dick, what in the - - - hell do you mean riding that horse like that? What did you come in for anyhow?

The kid stood confused and silent for a moment, but Smith pressed him for an answer.

"Well, Pete, dammit, the sky got all dark—an' if the World was really comin' to a' end, I dam sure didn't want to be left over at that camp by myself!"

This yarn supplied with the names of real cowboys of the old Bax X Ranch of Archer County has been told me more than once, but I still wonder if it was not a manufactured tale. Tales of this kind seem to float in the wind in the big ranch country. Possibly some of them are worth repeating and some of them are even based on authentic testimony.

Perhaps we had just as well poke around awhile in the land of big ranches for some more of this cow country gossip. The stories of thievery whether they deal with cow theft or not seem to head the list of the more dramatic part of this gossip.

Tom Smith (the name is fictitious if you please) was apparently an honest, upright West Texas cowboy until he helped to drive a trail herd to Montana. But something got in his blood in that far-away country and together with some confederates he took up the doubtful profession of train robbing. Perhaps the two other persons involved persuaded Tom to help with these holdups—that phase of their big time robbery business is not clear. At any rate some time after the three men had committed their first robbery, Tom became disgusted with himself and started back to Texas.

In New Mexico (or Colorado), on his way home he came through a severe snowstorm. Coming upon a young lady school teacher who was about to freeze to death, he offered assistance. He helped her to safety, and after making sure that she had recovered from the shock, he rode on to Texas. There was some correspondence between Tom and the girl but no reference to the Montana train robbery was ever made.

Tom went back to work at his old job on the ranch, but he was very nervous and kept a close lookout for any strange faces that might appear on his range. Rarely did he ever become separated from his pistols and he kept a good Winchester at the chuck wagon.

One day a party of men came to the wagon and began to inquire about Tom when he was not far away. He became suspicious and began to ride away from the wagon where his Winchester was stored, toward a herd of cattle that was grazing a few hundred yards from him. The strangers, who were in fact peace officers, felt sure from Tom's actions that he was their man and began to ride after him. A horse race followed with Tom not far enough in the lead to be out of range of Winchester fire. In desperation the fugitive cowboy rode straight into the middle of the herd, but a rifle ball struck his leg and caused Tom to realize that further resistance was futile. There was nothing left to do but surrender. Tom was convicted of train robbery and sent to the penitentiary, but he kept the whole proceedings from his school teacher friend.

However, the young lady who had escaped death in a snow storm through Tom's kindness finally learned of his predicament and set out to help him. At length she obtained a pardon for her erring cowboy friend, and though it may sound as if it is borrowed from fiction, she became Mrs. Tom Smith—proud of everything Tom had done except the Montana train robbery.

Most West Texas cowboys were honest as the day is long, but occasionally one of them went to rustling. Bill Jones (certainly that is not his real name) was the king among the cattle thieves who infested the country below the Caprock. Bill stole himself rich according to old cowboys, but everybody liked him even while the thievery was going on. He worked for one of the larger ranches for a short while, but easy money beckoned, and he answered the call. He set up his little ranch in some of the rough country near the big Matador range. Petty thievery was contemptible to Bill; so he went into the business on a larger scale. According to reports, he stole some eighty head his first year and then began to lay the corner stone of his ill gotten fortune. Two hundred head that should have borne the Matador brand were his by the end of the third year, and that was not half of his activities.

A man had to have land if he grew in the cattle business; so Bill did not neglect that part of the venture. He homesteaded some land in his own name and manuevered by various and sundry means to get others to help him acquire his grassland empire. Bill added a little high finance to his other successes and, in so doing, augmented his fast growing fortune. Certain kinds of script could be purchased for a song and applied to taxes at 100 cents on the dollar. He worked out a shady scheme to purchase script and pay other people's taxes with it, thereby in effect selling the discounted paper at face value. When death finally put Bill out of the cattle business, he was worth more than a half million dollars. Probably he had often felt himself pursued and turned quickly finding nothing in pursuit but his shadow, but he kept his stolen fortune to the last.

Bill had a neighbor whose name was not Jake Fry, but we'll call him that anyhow because there are no names foul enough. Jake was just a small time rustler but there may not be enough red hot coals of fire in all of hell to do him justice. It will be easy to understand what I mean in a minute.

One time when Jake had to stand trial for cattle theft he warned his own brother—whom everybody knew would tell the truth—not to appear as a witness. Before the trial Jake found his brother seated on a cotton wagon without a chance to defend himself. The contemptible cattle thief with gun belching fire advanced and fired until his helpless victim crumpled up and fell to the roadside. After that, Jake's neighbors placed him in their order of affections not much above a rattlesnake. But everybody knew that he was a man to be feared.

Several years earlier Jake had set the date of a roundup on his little ranch, and while he was about it, he sent out a warning for all Bar Z cowboys who didn't want trouble to stay away. He even called them by names—Doak Good the wagon boss and his brother Bob, Billy McIntosh, Charlie Hirsh, and Dave Ward. It's a pity we can't call the real names of this ranch and its top cowhands; a warning from a cattle thief was an everlasting testimonial to the good character of a cowboy.

Probably it was Doak Good that caused Jake's trigger finger to itch more than did all the other Bar Z cowpokes. Doak, a near neighbor of the cattle thief, had at one time refused to help the outlaw make bond.

"Don't you think that's presumin' on friendship a little too much to ask me to sign that bond, especially after you know I've laid out here in the hills for two or three weeks to catch you stealin' these cattle?"

Jake never did forgive Doak for this refusal and probably loaded his gun with revenge in mind on the morning of the roundup. Nobody doubted that the cow thief's little ranch was stocked with plenty of Bar Z cattle—and instead of timidly staying at home the Bar Z boys weren't the kind to back away from trouble.

When Jake warned these men of the old West to stay away he had just as well have issued an engraved invitation to each of them.

About the time this roundup of the motley herd of stolen cattle was well under way, the Bar Z boys rode up. Doak Good was in the lead. Jake made a quick motion for his gun, and Doak jumped from his horse with a 30-30 rifle in hand ready to fire. P-i-n-g—went a rifle ball aimed straight at the outlaw's heart and in a split second Doak was ready to pull the trigger again, but Bob yelled, "Don't shoot him any more, Doak; you've already killed him."

Somewhat regretfully the old cowhands of that time admit now that Jake lived over the rifle shot. Once when Dave Ward had taken on a little more than a man ought to drink, he told about the outlaw's wound.

"You could see his damned old heart beat through the hole that Doak put in him," was the way that Dave described it. But the bullet struck Jake just a hair's breadth above the heart and in due time he was back in the saddle ready to raid his neighbor's ranges again.

Perhaps a year or two later Dave Ward and Doak Good were riding toward a certain pasture gate; a few hundred yards across the fence and coming in their direction rode Jake Fry toward the same pasture gate. Doak, not expecting trouble, had left his gun at the chuck wagon. But now that he discovered his old enemy approaching he fully expected anything to happen. However, he rode unflinchingly toward the pasture gate as calmly as he would take a drink of water. Jake Fry without a halting step also rode toward the gate. Upon approaching each other, Dave Ward stepped off his horse, opened the gate wide, and led his horse back out of any possible line of gun fire. Not a word was spoken. Doak and Jake road silently past each other through the open gate. Then Doak stopped and sat still in the saddle with his back toward his outlaw enemy. Moments of anxiety seemed like hours as he awaited a death bullet in his back. Finally, still not turning around to look behind him, Doak broke his silence:

"Dave, is he over the hill yet?"

"Yes, Doak, I don't see him anywhere. I guess he's gone."

With the tension off, the two Bar Z cowboys rode ahead not knowing whether Jake Fry was also caught without a gun or whether he just didn't have the nerve to use it. Probably the outlaw never knew that Doak Good rode past him without so much as a toy pistol to use in his own defense. Both the Bar Z wagon boss and the cattle thief had observed an important part of the unwritten code of the old West—that a true rider of the range never runs from his enemy. Strangely, probably through fear, the cow thief had observed another part of that unwritten code when he failed to shoot his enemy in the back.

For a few more years Jake Fry continued to prey on the surrounding cattle ranches, but it became a trouble laden business. He seemed never to be more than two jumps ahead of the law and began to plan some schemes that might give him a little more security.

Once when the Bar Z's shipped a few cattle to market at Fort Worth, he slipped down there and bought them to ship right back to his little ranch along side the Bar Z. Thus he expected to be able to show that he owned some Bar Z cattle, that he had a legal right to keep. Who could distinguish between Bar Z cattle that he legally owned and any that he might choose to steal?

Jim Lord, the owner of the Bar Z Ranch, warned Jake not to

ship these cattle to the cow thief's own ranch. But Jake Fry was
not easy to scare. Brazenly, he warned the Bar Z cowboys not to
be present when he unloaded his newly purchased cattle at the
shipping point near his little ranch. He also warned Jim Lord
to stay away and even warned Dick Purdy who had often been
referred to as Jim Lord's body guard. Purdy was a good cowboy,
but he was also one of the expert gunmen of the West. He just
didn't feel at home without a little trouble now and then. In
Dick's older days he used to pat the fine gun that Jim Lord had
given him and say,

"Boys, this is the baby that I used to pick the teeth of those
bad men down at⸻."

Dick thrived on the prospect of a little gun play and no doubt
Jim Lord had an added sense of security when Dick was along.
The two of them set out to intercept Jake Fry's effort to unload
the shipment of cattle. They reached the little West Texas shipping point before Fry did and awaited his coming. While they
sat in a hotel lobby, Jake and one of his henchmen walked
brazenly past them into the wash room.

This time trouble didn't wait. Both Jim and Dick rushed to
the wash room. Without preliminaries guns began to bark and
as quick as you can tell it, Jake Fry lay dead upon the wash room
floor and his henchman had escaped by a back passage.

Jim Lord was tried for the killing, but the jury gave him an
acquittal. Probably any jury in the West would have done the
same—and secretly wished that there was some way to compli·
ment him for the shooting.

Many years have passed since this tragedy took place in the
little western hotel, but old timers still wonder if Dick Purdy didn't
actually kill Jake Fry and that Jim Lord as a magnanimous
gesture agreed to take the blame. Once Dick had boasted that he
had done the killing—but then Dick was drunk, and nobody was
sure what to believe.

During the long span of years from then till now, Jim Lord,
Dick Purdy, Doak Good and all the other old Bar Z cowboys have
died but legend carries on. It is still one of the unanswered questions in the gossip of the cow country "Who actually killed Jake
Fry?"

The old West with its rustlers was not the only direction to
look for cattle thieves. Some three or four years ago somebody

whisked away eight or nine cows and their calves that belonged to Lee Ribble, recently of Crowell, Texas. No clue has yet been found.

At the very beginning of this volume we have seen how it requires the everlasting watchfulness of the law and the cattle inspectors to keep up with the rustlers who prey on the big Matador range. Certainly none of the other great ranches are beyond the reach of thieves.

And so the modern phase of cattle rustling still goes on— truck loads of cows whisked away perhaps in the dark, or unbranded calves hauled in a one horse trailer. The thievery would most likely mount to large proportions except for the ever present vigil of the Texas and Southwestern Cattle Raisers Association. They see that the thieves pay the penalty pretty often.

And while we give credit to whom credit is due, the one time cowboy sheriff at Wichita Falls, the late Bob McFall should not be overlooked. As an officer of the law, Bob hunted the "cool off" pastures of the modern cattle thieves and ferreted out their dark schemes almost to the day of his death. He was typical of many such enforcement officials, but he added a little color to the job that most of the others missed.

All in all the cattle thieves of any age have been a very small percentage of the people in the cow business, but in the early day big ranch country there was on occasion enough gun play that accompanied the thievery to make the old West fairly sizzle with excitement.

But layng aside the gun play, the cow country also had its share of fun. There are those who would claim that a cowboy had a distorted sense of humor, but even that is a matter of view point. In truth there was little of the delicate type of humor in the land of cattle, but after all, cow work followed few of the rules laid down by Emily Post.

The stillness of the night in border towns was sometimes disrupted by staccato pistol music. The cowboy meant it in fun— his kind of fun. He was also likely to play pranks on those not initiated in the routine of the ranch. Sometimes he was down-right "onery."

Once a new arrival from Scotland came to the Matador Ranch. His knowledge of the cow business was limited to such things as he may have read, but none of his knowledge was gained from first hand experience. He was assigned to the remuda with the official title of horse wrangler. In short, he was placed in

charge of the cowboys' herd of saddle horses. On a large ranch
that was no small assignment.

One afternoon a division of Matador cowboys that worked
from a certan chuck wagon had moved their wagon a short distance under a hill without the knowledge of the Scotch horse
wrangler. This gave the Scotchman the double problem of keeping the horse herd together and of finding the new location. A
storm blew up with little rain but with lightning and thunder
that seemed to split the heavens wide open. Rallying all the skill
at his command, the Scotchman had rounded up his horses on a
hill top which by coincidence was near the new location of the
cowboys—although he saw neither cowboys nor wagon. A flash
of lightning and a sharp clap of thunder scared the horses until
they scattered in all directions like a covey of quail in a shinnery
patch. The onery cowboys waited in hiding at the foot of the hill
while the wrangler frantically tried to reassemble his herd. He
soon saw that his efforts were useless and began to yell for help.
By accident, the Scotchman rode up on the cowboys who inno-
cently engaged him in conversation, feigning not to know his
predicament.

"Where is the remuda?" one of the cowhands asked.

"When I came upon the divide the remuda fell apart," was
the excited reply. The cowboys restrained their laughter. One
of them held his face straight long enough to speak again.

"Why didn't you yell for help?"

"Three times I cried aloud, but no one answered," was the
wrangler's rejoinder. The cowboys rolled in laughter that was
altogether a mystery to the Scotch tenderfoot. In due time they
rounded up the horses but would have repeated the operation as
soon as the first opportunity arrived. In fact such oneryness on
the part of cowboys is limited only to the number of uninitiated
new arrivals on the ranch and the number of awkward predica-
ments in which the cowboys find them.

Another of the attributes of an old cowhand that defy
measurement was his love for his horse. In this regard there was
a certain kinship between the cowboy and the plains Indian. The
individual red man had been limited to a small spot of earth until
the arrival of the horse in North America. After the horse came,
such an Indian, as the Comanche, came gliding over the plains
like a fairy prince in seven league boots. With the horse he was
master of space.

The cowboy was similarly situated. Common chores that

separated him from his horse were especially distasteful. To repair a fence or to grease a windmill was a little beneath his dignity, and it has not been too easy to change his view point. To ride with a swinging loop at breakneck speed and bring an unruly yearling under control—that was life to a cowboy.

Not long after 1880, a cowboy in West Knox County was assigned the task of helping to dig a water well. He had been at the job several days when a neighbor rode up with a newspaper in his hand. The visitor dismounted and began to read from the newspaper. The cowboy was at work in the bottom of the well and could barely hear the various news items as they were read. Finally, the visitor read about a new bounty that had just been placed on certain kinds of obnoxious wild game. For instance, a prairie dog's scalp would bring three cents and the scalp of a coyote would bring three dollars. At this last bit of news the cowboy stopped digging and called out from the bottom of the well.

"Read that over again, will you, pardner."

The visitor reread the item and the cowboy began to imagine himself riding the range on horseback collecting the scalps of these predatory animals and cashing in for a lot of easy money. Again he called out from the bottom of the well.

"Draw me out of here and let some pore son-of-a - - - have this job."

It was the cowboy's way of resigning from a job that was beneath the dignity of a skillful horseman.

It is not just idle talk that a cowboy can often remember a horse as long as he can remember a man. Jim Olds, late of Crowell, Texas, who used to work for the old R2 Ranch that once covered the range around present-day Chillicothe remembered a number of old cow horses. "Joe Johnson was manager of the ranch sixty years ago," Jim told me. "He rode old Dick, a smooth bay horse, and some of the time he rode a fine white horse that wore a triangle brand. Joe rode a black horse part of the time too."

Jim shifted in his chair a little and told how the horses that pulled the chuck wagon, old Bugle and his yellow colored team-mate, used to hang around the wagon and eat biscuits while a gallery of cowboys looked on approvingly.

Feeding pet horses at the chuck wagon was not at all limited to the R2 Ranch. Genial, prank-loving Billy Pressley who spent a quarter of a century on the Pitchfork Ranch fed his favorite black horse biscuits from the "Fork" chuck wagon on many occasions. So did Jeff Harkey, an early wagon boss for the "Matadors," sympathetically administer to his pets with chuck wagon biscuits. Old House and Joiner were the two that Jeff fed out of food provided by non-resident Scotch stock holders—but he probably would have fed these two old pets in front of the board of directors of his company had occasion presented itself.

But the bond between a cowboy and his horse covered a wider range of feeling than that of a family who love a household pet. In addition to love, there was admiration almost akin to hero worship for the horse that knew how to cooperate fully with his rider in doing the roping chores, and that admiration was even stronger for the cutting horse that could turn and twist as much as needed to cut an unwilling yearling out of a herd.

Sometimes the horse did his work so well that he embarrassed the rider. Charley Benson of the old Two Circle Bar, whom J. T. Bond of Jayton described as "the best cowboy I ever knew," came out of one episode with a much ruffled appearance and a red face. When he and Tom Davis tried to round up some horses, old Robert, his cutting horse, with a quick turn after an offending horse, cut from under Charley and left him sprawled out on the ground. This ace workman among the Two Circle Bar cowhands brushed himself a bit, charged himself with the error, and admired the efficiency of his favorite horse all the more.

The implicit faith that both Sam Graves and Bud Arnett had in old Hub has already been told, but perhaps the medal should go to Ed Carnes of Wichita Falls for about the highest degree of confidence that a man ever had in a horse. Ed rode a fine 1100 pound sorrel that belonged to his employer, Tom Jones. Whatever Ed threw the loop of his lariat around, he expected the sorrel to stop dead in his tracks. Many a time his horse had been forced to drive both front feet in the ground and to set back like a stubborn donkey, but always, the animal at the other end of the rope came to a sudden stop or flopped in the dirt. The horse had never failed, and Ed believed him capable of stopping just about anything that moved.

One day after Ed may have had a little stimulant, he and the sorrel rode out near the coal chute in the southeast part of Wichita Falls, perhaps seeking adventure . . . and adventure in the form of a moving switch engine lay just ahead of them. Ed put spurs to his mount and loped toward the engine and then, with a great whirl of his lariat, he threw a loop over the smokestack. His 1100 pound sorrel horse set his feet in the ground ready to bring the thing to a halt, but this time something went wrong. However, before any real damage was done, the engineer who knew Ed stopped the engine and saved the day for both horse and rider. The sudden stop of the engine had probably prevented a crash with many broken bones in the bargain, but it had also saved Ed Carnes' confidence in a horse that could stop just about anything that moved.

But the gossip column runs into many pages. Let us return to the journey through the big ranches.

11—Inside the Pitchfork Gate

After this short detour into the gossip of the cow country, let us go back to Dickens, the little cross road village just below the foot of the Plains. Here just seven or eight miles east of the Caprock let us resume our journey among Texas ranches.

I continued eastward along the same fine paved highway that I had previously followed from Lubbock. The countryside is no longer the endless flat prairie that is found everywhere on the High Plains. Here I passed over a succession of red hills that are separated by the many little head branches of Crotan Creek. The highway cuts across an arm of the Matador Ranch here for several miles.

The first movement of cattlemen to this area came just before 1880. C. L. (Kit) Carter, then president of the Cattle Raisers Association, accompanied by a number of his associates came on a prospecting trip[1] to this country below the Caprock in 1879. First, he and his party went to Blanco Canyon, but they evidently found most of the choice locations there already occupied. His party then traveled across Dickens County and chose a range on the head of the north Wichita River in and adjoining the northeast corner of Dickens County. Carter soon built a small box house and continued to ranch there until 1887. The area grazed

1. Fort Griffin Echo. August 2 and November 1, 1879.

by Carter and his associates was long known as the Ross range, the name originating from Carter's "R O S" brand.

My highway soon passed out of the Matador into the Pitchfork Ranch. Some twenty-two miles east of Dickens the very modern headquarters of this Pitchfork property lie a few hundred yards south of the road. The buildings were built fifteen or twenty years ago, but the ranch itself is more than seventy years old.

Even before the present Pitchfork organization was formed, Jerry Savage from whom the Pitchfork brand was purchased was ranching on the head of the South Wichita River. The news columns of the *Fort Griffin Echo* mentioned this Savage Ranch on October, 1880, and again in January, 1881. From these news items there is little room to doubt that the Pitchfork brand was in Dickens and King Counties seventy-four years ago.

In searching the records to find the origin of the Pitchfork brand, we note that such a brand was registered in the name of Norman Savage in Matagorda County in 1843 while Texas was still a republic.[2] His old Pitchfork brand faced downward and had no middle prong, while the present-day brand faces upward and has added a middle prong. Is it possible that Jerry was a descendant of Norman Savage and merely made a new adaptation of the old family brand?

But regardless of the possibility of kinship between these men whose family names were the same and whose brands were so nearly alike, we can be sure that the Pitchfork brand was one of the first that came to the West Texas open range, and we can know with certainty that it has continued to grow in importance all through the years.

Perhaps the real origin of the Pitchfork Ranch was the friendship that existed between J. S. Godwin and D. B. Gardner. Godwin had once ranched above the site of the new Wichita Falls Lake Kickapoo, on a stream that is still known as Godwin Creek; but sold his cattle and soon became associated with Gardner in the Fort Griffin country. Sometime in 1881 the two men bought the Pitchfork brand of cattle from Jerry Savage.[3] The "Centerfire Dugout" went with the deal but there was no land ownership

2. Gus L. Ford, Texas Cattle Brands, 11.
3. Margaret Elliot, "History of D. B. Gardner's Pitchfork Ranch of Texas," The Panhandle-Plains Historical Review, XVIII, 31. In this article Miss Elliot refers to the firm who sold the Pitchfork Cattle as Savage and Powell. The old Fort Griffin Echo of October 2, 1880 and January 15, 1881 mentions Jerry Savage as owner of the ranch without any reference to a partner. Certainly the discrepancy is minor and it is easily possible that the old newspaper, though published at the time of these events, may have erred.

involved. Late in 1881, Godwin sold his interest to **Eugene F. Williams** of Saint Louis.⁴ Williams was associated with men of means and his connections most likely made the expansion of the ranching venture an immediate possibility. **A. P. Bush, Jr.** of Saint Louis bought an interest with Gardner and Williams in 1883, and late in December of the same year, the Pitchfork Land and Cattle Company was organized at St. Louis, where the head office of the Company was established, and where it has continued until today. Samuel Lazarus of Sherman, Texas, sold the new company 25,000 acres of land, 3750 head of cattle, and leases on additional grass lands for $125,000. He took stock in the company for most of his sale price. Gardner, Williams, and Bush transferred to the company their cattle, horses, and other equipment which they had acquired for a $150,000 block of stock. A. D. Brown of St. Louis bought $30,000 worth of stock and became the first president of the company. Some land owned by Williams and grass leases owned by Williams and Gardner also became the property of the company.⁵ The company was capitalized at $300,000 with all but $5000 of the stock issued.

One of the most important things the company did at their first meeting was to elect D. B. Gardner as manager. Gardner held that important position for the remaining forty-five years of his life and made many thousands of dollars in dividends and built up a large surplus in the company's treasury. But Gardner, during those forty-five years, did far more than to perform just the bare duties that resulted in successful management. He was so very human with his entire organization and with people in general that he has come to be known as "the grand old man of the Pitchforks."

He began unconsciously to earn that title from the very beginning. One day a thirteen-year-old orphan boy by the name of Billy Parks came to the ranch. The boy had no connections and in fact did not correctly know his own name. Actually on later investigation it was learned that he was really Billy Partlow. But the name did not matter, for Gardner put him to work on the ranch and he became known far and wide as the "Pitchfork

---

4. **Fort Griffin Echo,** December 10, 1881.

5. Margaret Elliot, **op cit.,** 36-37. The article by Miss Elliot in **The Panhandle-Plains Historical Review** of 1945 tells not only the details of organization of the Pitchfork Ranch but gives the details of the ranch history far more completely than can be attempted here. Much of the remainder of this chapter is based on Miss Elliot's excellent article, which fills sixty-seven pages (pp. 12-78) of the above publication.

Kid." The name stayed with him for the rest of his life even though in later years he worked for the Matador.

Gardner must have had a "way with boys." Press Goen came to the ranch when he was only fourteen years old, and Billy Pressley when he was only thirteen—but both boys became valuable men and paid back great dividends to the ranch in services. Press was wagon boss for fourteen years and part of the time Billy was his able assistant. Both boys availed themselves of the privilege granted by Gardner of building up a herd of their own, and finally both men began successful independent ranching careers. A seemingly little thing that Gardner did for Pressley when he left the ranch was to allow him to keep Chub, his favorite horse, but the act was a finer character testimonial for Gardner than a whole stack of written recommendations would have been.

At the end of Gardner's long career as ranch manager, the "Pitchforks" owned 120,000 acres of land stocked with cattle and a large number of man-made watering tanks and something more than thirty windmills to serve the cattle. The land was fenced into three large pastures and, all in all, was so organized that it paid good dividends.

However, there were still phases of the ranch make-up that could stand improvement. Probably Gardner would have increased the cultivated area and with it the feed production had it not been for the fact that cowboys in general refused to do farm work. He lost some good men in 1906 because he put them to pitching hay, and most likely he hesitated to aggravate this sore spot with his cowboys. Nevertheless a modern ranch needed to supplement grazing with outright feeding in order to help the cattle stand the severe winters. Such procedure would decrease the death rate among cattle during blizzards and add valuable pounds of flesh to the animals while so doing.

Some two years after Gardner's death, a modernization program was inaugurated under Virgil Parr as manager. Parr added wells and windmills until the ranch had a total of sixty-five and until no cow on the ranch had to walk more than a mile and a half from grass to water. He increased the cultivated land to about 4000 acres, dug great trench silos,[6] and supplemented the pasture grasses with much fat-producing food.

Just imagine a fellow doing that in a ranching country long steeped in traditions against farming! Imagine the conversations

6. R. Ernest Lee of Wichita Falls, Texas, was the engineer employed by Parr to work out the details of these silos. Mr. Lee is the source of this information.

of old time cowboys who stood around the store fronts and filling stations in the little towns of Guthrie and Dickens! Yet Parr could take out his slide rule and show how it all paid dividends— but the slide rule itself was an object of local prejudice. What business did a cattleman have with such a stick? Parr, in cow country gossip circles, most likely would have become known as a slide rule cowboy, had any body known the name of that ivory covered piece of wood that he carried around with him.

But the time had come for science to take a hand in the production of cowflesh. Millions of acres of grass had gone under the plow and the American public might wake up one of these days on half rations of beefsteak! Would these methods instituted by Parr at the Pitchforks, if emulated by the whole cattle country, make up the deficiency? The answer to that question is indeed a big order and should require an understanding of a still greater section of the cattle country than we have thus far examined.

But it is time to sidestep this important question for a while and examine first-hand the fine headquarters of this much-talked-of Pitchfork Ranch.

I drove in at the Pitchfork gate and glanced hastily at the half dozen or more buildings that made up the ranch headquarters. The Bunk house had the appearance of a rather neat but modest school dormitory. Here each cowboy was assigned his own individual room quite different from some preconceived notions about a modern ranch. The mess hall half hidden in a grove of trees, was next to draw my attention. It was somewhat distinctive with its steam tables and other modern equipment designed to see that Pitchfork cowboys are well fed. The manager's home in the deep shade of shapely trees with its several white gables gave the appearance of luxury and roominess.

D. Burns, the manager, was not at the ranch on this particular day, but two of his men—Coy Drennan, the wagon boss, and Riley Thacker[7], bronc buster deluxe—extended the courtesies of the ranch in most hospitable fashion.

Riley will most likely give me a gentle reprimand when he reads about himself as a great bronc buster. But cowboys never rate themselves at full value, and according to the general opinion, Riley can stay on a horse's back when the going gets tough.

---

7. This is the same Riley Thacker who was mentioned in an earlier chapter as a cowboy with the 6666 Ranch. As a matter of fact my day's visit on the 6666 Ranch came later than my visit to the Pitchforks. In the meantime Thacker had become an employee of the 6666 outfit. This is one of the few instances in which my journey did not occur in the exact order mentioned in this account.

But even this is an unfair way to rate Riley's duties with the Pitchfork Ranch. I asked him if he did not have plenty of horses that made him do "some pretty hard riding."

"Not if I can talk 'em out of it," was his quick comeback. "If you teach a horse to pitch, he'll pitch when you don't want him to."

That kind of talk was a surprise to me, but gradually I began to understand what it meant. The old fashioned method of turning a young colt out on grass to kick up his heels in utter freedom until he is two years old and then to manhandle him into submission is not the Pitchfork idea. As soon as a colt is weaned from his mother he is fed and led around by a halter at intervals and ridden a short distance as soon as his little body can, without injury, support the weight of a man. Thus by gentle and easy methods he is trained to be a cow pony. Most horses so trained never reach the pitching stage and make far more useful mounts than those that are allowed to become wild and then ridden to unwilling tameness by a bronc buster. Riley Thacker is not primarily a bronc buster—he is a trainer of young horses. He can "ride 'em" if it takes that, but he is far better satisfied with a horse that never learns to pitch.

What is said here about the training of horses at the Pitchforks might just as well have been said about the 6666 Ranch. Down at Guthrie, George Humphries has a regular stable where young colts are fed and trained. A number of ranches have learned that bronc busting with all its dramatic aspects does not pay dividends as well as the gentler methods of training young horses.

Before I left this interesting ranch headquarters, both Drennan and Thacker had opened up their collections of ranch country photographs and, in keeping with the usual generosity known throughout the cow country, had permitted me to make use of any that suited my purpose. Thus, with whole envelopes stuffed with life-like scenes from the daily work of the ranch, I drove back out the big gate to the public thoroughfare.

12—A Buffalo Hunter Camped Here

This trip to the Pitchfork Ranch, although on the main Lubbock to Seymour highway and to one of the important present-day ranches, was made only by turning off the base route that was mapped for this journey through the old ranching country. The 6666 Ranch headquarters, it will be remembered, is in Guthrie which is only eleven miles east of the "Pitchforks." Since this trip has already included the 6666 Ranch, it becomes necessary to turn around here at these Pitchfork administrative buildings and travel the twenty-two miles back to Dickens.

In these twenty-two miles, earlier in the day, I had crossed a stem of the great Matador Ranch. On this return to Dickens, while again crossing this arm of that giant ranch, I looked southward toward the Croton Breaks. This chopped-up patch on the map of West Texas, not very plainly visible from my road, undoubtedly would take first prize among the bad lands of the South Plains.

In one rather ludicrous particular it out-does Bryce's Canyon which a modern magazine writer describes as "one hell of a place to lose a cow." The Croton Breaks, though picturesque in their own right, cannot compare from a scenic viewpoint with magnificent Bryce's Canyon, but they are even a worse place to lose a cow. The Matadors carefully rounded up their cattle from these bad lands in 1936. Some 600 head were driven from this fluted piece of earth.[1] Among them were many six-year-old bulls and cows that had never seen a branding iron—Mavericks in the original sense of the word! But even more unbelievable than these curiosities of the modern day, there were at least 100 steers that had long run wild like animals in the jungle. Some of them bore an old Matador date brand known to be seventeen years old!

Since 1950 a new chapter has been added to the history of the Matador Ranch. It will be duly reported later in this volume. Meanwhile let us return to this journey which was made a little before 1950.

At Dickens I turned north along the paved highway toward Matador. At the time this journey was made, Press Goens, the fourteen-year-old boy who helped to make Pitchfork history, lived a few miles west of this road and some seven or eight miles from Dickens. Press had become an important, independent cattleman with a dozen or more square miles of grasslands in his own ranch.

Eight or nine miles north of Dickens, the ranch home of the late Bill Stafford fronts directly on the highway. Bill was probably the last living cowboy who worked for the Spur Ranch before the Scotch syndicate took over. The beginning of his employment with the Spurs dates back to 1883.

I parked at Bill's gate and had a nice visit with him. Drouth was causing cattlemen and farmers some worry at the time; but Bill, who had spent more than sixty years out here below the Caprock, discounted the dry weather talk with some interesting comment. "I know it's dry," he said, "but people forget that a good rain and three days' time can make it look like a new country." Soon the conversation drifted to old trails and other matters not so closely connected with the cattle business and at length I reluctantly went on my way little dreaming that it would be my last visit with Bill Stafford.

My highway led northward toward the headwaters of the Pease River. It was only a few miles to the south branch of the stream known as the Tongue River. A good bridge takes the traveler high above the sandy bed of this water course. There is a

1. Minnie Timms Harper and George Dewey Harper, **Old Ranches**, Dealey and Lowe, Dallas, Texas, 1936, p. 96. The Matador owners have sold more than 40,000 acres of these bad lands to the Pitchfork owners. A more accurate accounting of acreage will be made in a later chapter.

stream of excellent spring water that in dry times sinks down in the sand along this stretch of river. A mile or perhaps a little more upstream, the famous Roaring Springs break out of the rock formation a hundred yards from the river. The incessant roar of the water as it pours over ledges of rock gave the springs the name. The Matador Cattle Company, owners of this old watering place, built a concrete swimming pool just below it to the delight of the many bathers who frequent the place.

Stone metates on which Indians once ground their corn and other similar relics in stone have been found here. It was doubtless one of the great gathering points for red men long before written history began.

There are still other springs that have in ages past broken out across the area just a few miles below the Caprock, but none of them seem to have quite the significance of Roaring Springs. However, so far as the story of white men is concerned, a less conspicuous watering place twelve miles north of this famous camping spot has played a more important role.

In about 1877, a buffalo hunter whose name was Ballard selected as his headquarters a spring a mile southwest of the present-day town of Matador.[2] At this spot and near it Ballard collected the skins of many buffalo as the great Texas hunt thinned out the last of these massive beasts. At the end of the hunt, Ballard left the spring and Joe Browning moved in.

Browning held the place for a short time, while at intervals a man destined to possess one of the well-known names of the west looked longingly at the spring and surrounding range. The interested stranger was none other than H. H. Campbell who would perhaps deserve a place among the dozen greatest ranch managers in Plains history. Campbell bought Joe's interest[3] and began to assemble one of the immense ranches of the West with his headquarters at Ballard Springs.

By 1879, he together with A. M. Brittain, Fort Worth banker, had assembled $50,000 and had begun to make purchases and take options.[4] The new company was given the colorful name, Matador, which gained a permanent place in Texas geographical names. At first they began to brand their cattle with a 50 M, but late in December the rather large Dawson herd was purchased and convenience dictated a change in this respect. The Dawson

---

2. Ibid, 84.
3. Ibid, 84.
4. Laura V. Hamner, Short Grass & Longhorns, University of Oklahoma Press, Norman, Oklahoma, 1943, p. 149.

cattle from South Texas branded with an odd shaped V were some 1500 strong. Why not adopt the Dawson V? This is exactly what they did[5]—and now three quarters of a century later the present owners of the Matador cattle still brand with the Dawson V.

Soon, perhaps in 1881, the company acquired some 8000 head from Bob Wiley and Tom Coggins[6] who had trailed them across the Plains from the Pecos River. These cattle were part of the famous Jingle-Bob herd which John Chisum had disposed of as part of a vain effort to maintain the possession of his widely known Jingle-Bob Empire. But the sun that was slowly sinking on the cattle king of the Pecos was beginning to rise on Henry Campbell and his associates on the upper Pease River. In January, 1881, the new concern bought 5000 cattle from Tobe Odom[7] of Fort Chadbourn, Texas; and by various other transactions the expansion of the new ranch continued.

But the crowning event in early Matador history came in 1883 when banker A. M. Brittain went abroad. He came in contact with some Scotch financiers at Dundee, Scotland,[8] and raised $1,250,000 to carry on the expanded plans. Campbell was retained as manager of the ranch until 1891 when he resigned to go into the ranching business for himself, but not until he had seen the new company far along its way. The Matadors were often branding more than 20,000 calves per year before Campbell resigned.

In fact the ranch, until recently in the hands of the same Scotch syndicate, has grown to be one of the greatest cattle raising organizations on earth. The ranch that centers at Ballard Springs where the old buffalo hunter once held forth at most covered more than 450,000 acres.[9] From the Crotan Brakes southeast of Dickens to the extreme northwest corner of the range is about sixty miles and another line across the ranch from the edge of the Staked Plains west of Matador to the extreme edge of the

5. Interview with Harry Campbell of Matador, Texas, June 15, 1954.
6. With some Durham blood these Jingle-Bob (Fence rail branded) cattle were a better herd than the V branded Dawson Longhorns—so says Harry Campbell, John Dawson seems to have bought an interest in these Jingle-Bob cows before the sale to the Matadors, but the herd still belonged to Wiley and Coggins in the spring of 1880 after the V brand had been adopted. See the **Fort Griffin Echo**, for February 7 and March 6, 1880.
7. **Fort Griffin Echo**, January 22, 1881.
8. Laura V. Hamner, **Short Grass & Longhorns**, 150: Minnie Timms Harper and George Dewey Harper, **Old Ranches**, 87.
9. Some 40,000 acres recently sold to the Pitchfork Ranch owner reduces the size of this ranch to some 400.000 acres. A later change in ownership will be mentioned in due time.

east pasture would measure forty miles or more. A mammoth range indeed—but it was not much more than half of the full grown Matador Ranch. The company acquired 270,000 acres south of the Canadian River above Amarillo and another block of 120,000 acres across the river.[10] There were also some 17,000 acres in Montana that belonged to this company of Scotchmen, and this latter property was so located that almost limitless free grass adjoining was available without formalities. Such an immense acreage prompts us to ask, "Had the Matadors become the largest ranch in the United States?" It is the intention to go into that matter in a later chapter, but it will not be improper even here to cast a glance backward for a moment and view some of the greatest ranches of the past.

Often since about 1870, some large cattleman (or some large ranching organization) has run his holdings up to near 1,000,000 acres of land or to perhaps 100,000 cattle. Rarely has any ranch or ranches ever equaled or exceeded those figures. Some estimates of doubtful accuracy place the herds of L. B. Harris of Atascosa County above 100,000 cattle soon after 1870. If the estimates are correct, he was possibly the largest owner of cattle in his day. But if so, his leadership was not of long duration, for that lead has changed back and forth over a span of years like a see-saw horse race. John Chisum whose cattle empire once stretched out over 1,000,000 acres of free grass along the Pecos River[11] is said by some to have approached the 100,000 cattle mark a few years before 1880, but by the end of that decade his curve of greatness was fast going down hill. Chisum died in 1884 with his fortune greatly reduced.

It is possible that the Prairie Land and Cattle Company, a Scotch syndicate, grew at the right time to move into first place just as John Chisum began to fade out of the picture. By 1881, these Scotchmen had purchased the great herds that had belonged to Jim Hall's two brothers in northeast New Mexico; they had also purchased the cattle that covered a great slice of southeast Colorado, completing a veritable empire of cows and grasslands that spread well into two states. But the Scotchmen did not stop until they had bought up nearly everything on four feet

10. A more accurate listing of acreage is given in a later chapter.
11. Most accounts of John Chisum refer to him as a great cowman: see J. Evetts Haley, George W. Littlefield Texan, University of Oklahoma Press, Norman, Oklahoma, 1943, pp. 137, 149. The booklet Old Ranches by Minnie Timms Harper and George Dewey Harper (pp. 63, 64) places his Pecos River range at more than 1,000,000 acres and his herds at 80,000 to 100,000 cattle. In his book, Cattle Kings of Texas, C. L. Douglas (p. 121) credits Chisum with 100,000 cattle.

along the Canadian River north and northwest of the site of present day Amarillo, Texas. Colonel Littlefield sold some 12,000 head for a cold quarter of a million dollars and other nearby cattlemen joined in the mass sale of cow flesh. Before the end of 1881, the "Prairie" had run its ownership to near 100,000 head[12] of Texas, New Mexico, and Colorado cattle and perhaps for a brief moment had scaled the dizzy heights to first place among American ranches—but their day of eminence was indeed brief.

Next to nose upward among the cattle kings was Colonel Goodnight with the aid of his financial godfather, John Adair. He built up a mammoth ranch in and near Palo Duro Canyon some twenty miles southeast of Amarillo. He just touched the magic figure of 100,000 cows and including leased land he grazed the great total of 1,335,000 acres.[13] But if he reached the lead in the race for first position among American ranches, he was soon outrun for the very odd reason that Texas just then needed a capitol building. The state traded a little more than 3,000,000 acres of land to outside capitalists to erect that building and the new owners of the 3,000,000 acres decided to convert their vast holdings into an immense ranch. The widely known X I T Ranch was the result. By 1887, it was stocked with 150,000 head of cattle[14] and had in the matter of magnitude shattered all records among American ranches.

Also in 1887, when the X I T Ranch had skyrocketed so far above all others, Colonel Goodnight and Mrs. Adair (for John Adair had died in 1885) divided their holdings. Thus the large ranching venture begun by Colonel Goodnight was somewhat reduced, although the holdings of Mrs. Adair were still very large.

Significant in the long time race for leadership, although at the time far behind the X I T, were at least five other ranches. The well-known Captain Richard King in the part of Texas below Corpus Christi had built up a half million acre ranch at the time of his death in 1885.[15] Mrs. King with her son-in-law Robert J. Kleberg as manager, slowly expanded the ranch over a long period of years. The ultimate outcome of that expansion will furnish some of the color for a later chapter.

While the King Ranch of south Texas was expanding, so was

12. C. L. Douglas, **Cattle Kings of Texas**, Cecil Baugh, Dallas, Texas, 1939, p. 297.

13. J. Evetts Haley, **Charles Goodnight Cowman & Plainsman**, Houghton Mifflin Company, New York, 1936, p. 325.

14. J. Evetts Haley, **The XIT Ranch of Texas**, 89.

15. Fortune (Magazine) December, 1933, 103; Gus L. Ford, **Texas Cattle Brands,** 148.

the vast spread of grasslands and cattle of Dan Waggoner—the same cattleman who had once rounded up his herds at the site of present-day Wichita Falls. Without repeating here the details of Waggoner's expansion, it may be said that by 1890 or a little later, he could count among his possessions some 80,000 or 90,000 cows.[16] He was ready to threaten the lead if opportunity presented itself.

Colonel C. C. Slaughter was another of the giant Texas cattlemen who might, with his great spread of acreage and cattle north of Big Spring, Texas, have threatened the lead[17] in his line of endeavor had it not been for the X I T . Also the Cunard interests cut a great figure in the cattle world, with their Francklyn Land and Cattle Company[18] in the Texas Panhandle eastward of Amarillo and in old Greer County in the southwest part of what is now Oklahoma. Their venture in the cow business both grew and died like a mushroom with only two or three similar businesses above them.

While the Cunard interests were thus building their great house of cards, the unique cattle baron, Shanghai Pierce, was fast rising to top position among the cowmen of the Texas coast. By 1888, just a little too late to threaten the national lead Shanghai and his brother had purchased 1,500,000 acres of grasslands[19] near the Gulf of Mexico. Perhaps there were other ranches of first magnitude that should be considered here[20], but these suffice to show the immense proportions attained by the leading cattlemen.

Along with the growth of all of these ranches was the slow steady progress of the Scotchman from over the sea who had undertaken to build up a cattle empire at and around the one time buffalo hunters' camp at Ballard Springs. They branded 25,000 calves in 1887[21] and soon leased additional acreage in the

16. An interview with Jim Marlow (a pioneer of Wichita Falls, Texas, who was familiar with Waggoner's operations of that early period.) See Gus L. Ford, **Texas Cattle Brands, 148**. The estimate from Ford's book places the number of Waggoner's cattle marketed per year (in the 1890's) at 40,000. If correct, the Waggoner herds must at that time have exceeded 100,000 head.

17. The holdings of C. C. Slaughter are described by C. L. Douglas in his **Cattle Kings of Texas** (p. 173) as covering an area of fifty by eighty miles. In another publication Slaughter is credited (in 1905) with the immense total of 100,000 cattle and 1,500,000 acres of land (both owned and leased). See Gus L. Ford, **Texas Cattle Brands, 137**.

18. The Francklyn Land and Cattle Co. is credited by Laura V. Hamner in her book, **Short Grass & Longhorns** (p. 231) with from 70,000 to 100,000 cattle and 700,000 acres of land owned outright and control over 1000 sections.

19. C. L. Douglas, **Cattle Kings of Texas, 152**.

20. Probably the Swan Cattle Co. (A Scotch concern that in early days began ranching in Montana) should be in this list of the near great among old ranches.

21. Laura V. Hamner, **Short Grass & Longhorns, 151**.

Panhandle northeast of Amarillo. In 1902, this company of Scotch investors, properly called the Matador Land and Cattle Company, purchased nearly 200,000 acres in Oldham County along the Canadian River to the west of Amarillo. Now the company had two-thirds of a million acres and were ready to move upward at every good (and financially sound) opportunity.

This last purchase of 200,000 acres held a double significance in the race for American leadership in ranching, for the land was purchased from the X I T Ranch itself. This 3,000,000 acre giant was beginning to sell its lands! In fact we have already seen how that sale of lands actually began in 1901. By 1906, some 2,000,000 acres[22] of the once vast X I T holdings had been sold, and by 1912 the last herd of cattle belonging to this mammoth ranch had passed to other hands.[23] Thus the king among cattle ranches had abdicated his throne. The race was now left to the Kings, the Waggoners, the Slaughters, and the Matadors.

In 1904, the Matador company secured a grass lease on some 500,000 acres of the Cheyenne Indian Reservation[24] on the Missouri River in South Dakota, and at about the same time they also obtained a lease along the Saskatchewan River in Canada. Obviously by 1906, when the X I T company had disposed of two-thirds of its land, the Matadors with well above 1,000,000 acres of both deeded and leased land had arrived at the lead so far as cattle range is concerned. In 1913, these aggressive Scotchmen leased the Belknap Indian Reservation near Harlem, Montana—a 550,000 acre addition to their range. The Matadors were now in striking distance of 2,000,000 acres.

In a year or two, however, the Cheyenne 500,000 acre lease was relinquished but the Coburn Ranch of nearly 20,000 acres, which included almost unlimited free range adjoining, was bought in its place. This new purchase was beside the Belknap lease. Some eighty-five miles east of this purchase the Matadors bought still more land. This time it was the DeRicqles Ranch with a free range adjoinnig it that was even greater than the area once grazed by John Chisum's Jingle-Bob cattle.

The Matador cattle interests soon gave up their Canadian lease but replaced it by a lease on the Pine Ridge Indian Reservation in South Dakota.

This last lease they gave up in 1922 after holding it two years

22.   J. Evett Haley, The XIT Ranch of Texas, 222-3.
23.   Ibid, 220.
24.   For an account of the holdings of the Matador Ranch out of Texas see Minnie Timms Harper and George Dewey Harper, Old Ranches, 97-8.

but continued to hold the 550,000 acre Belknap lease in Montana and the Coburn and DeRicqles ranches that were near by. Back in Texas, the Matadors later purchased some 70,000 acres beside their Oldham County ranch and 120,000 acres adjoining on the north. Thus with some 853,000 acres[25] of land which they owned outright and the much larger leased and free range in Montana they had presumably become the greatest American ranch after the decline of the X I T. Even so, their leadership is often disputed. Let us discuss it in a later chapter.

I crossed the Tongue River and turned off the pavement and up the river a little more than a mile to my left to the Roaring Springs. This short side trip took me along a partially improved road for perhaps a mile and into the Matador pasture for several hundred yards. I parked the car on top of a small rocky hill and scampered down the steep slope westward to the spring. Pouring noisily out of the rock ledges north of me, this never ending stream of living water ran swiftly along a spacious concrete ditch beside the fine swimming pool which the Matador Ranch management had provided. But the pool was empty! The place was without bathers for it was the wrong season of the year.

I sat down on the concrete edge of the pool and watched the endless pouring of the never-tiring spring. It was not very different one hundred, two hundred, or even a thousand years ago! Red men had made it a meeting place for ages past. Men of different tongues had come together here—some of them from across the Plains. This fact of the mingling of people of different languages is said to have been the origin of the name "Tongue River"—the stream beside the spring.

If one could have hidden himself here and secretly watched and listened to this human story of the past, he could now fill many of the blank pages in the history of the West. The experience doubtless would have been packed with many thrills. But if some prehistoric red man could come back to life and secrete himself beside the pool when the bathing season is in full swing, he too could most likely have a story punctuated with exclamation points.

This is only one of the numerous streams of water that break out of the ground in this belt that lies along the east edge of the

25. John Mackenzie to J. W. Williams, August 25, 1947. The correct area was 853,128 acres on January 1, 1947. More than 40,000 acres have been sold to the Pitchfork Land and Cattle Co. since that date. Later the entire ranch was sold and the new owners sold parts of it to others—but the account will be given in a later chapter.

Caprock, but this spring is probably the most prolific of them all.

I climbed the hill to the car and drove back to the highway again near the north end of the Tongue River bridge. Here I resumed my journey northward. It was three miles to the town of Roaring Springs and only a fraction of a mile beyond to Dutchman River, another head branch of the Pease. There are some farms in this community around the town of Roaring Springs, but much of the back country (as is the case for thirty miles to the southeast) was the almost endless stretch of grazing lands of the Matador Ranch.

Some seven miles on the highway from the Dutchman River was the turn off to the Matador headquarters. Several hundred yards west of the highway were the company offices near the same spot where Ballard, the buffalo hunter, once held forth. Seventy-five years ago H. H. Campbell came here and set up the beginning of the great ranch. His headquarters then were nothing but a dugout. A year later Mrs. Campbell came out from Waxahachie, Texas. She objected to the dugout and a house was built of lumber. From then till now, the headquarters have grown with the years—always located at Ballard Springs.

In 1891, the Campbells left the ranch and A. J. Lingertwood became local manager with Murdo MacKenzie as superintendent. Following Lingertwood, J. M. Jackson was manager from 1909 to 1924. In later years M. J. Reilly was manager until his death in 1946.

I turned off the highway to visit the ranch headquarters. It was something near a half mile from the pavement. A number of neatly finished buildings were grouped along the draw eastward of the spring where Ballard, the buffalo hunter, had camped. To the north at the top of the hill was the manager's home with its long spacious porches and its well clipped lawn and wealth of shade trees. It was Mr. John V. Stevens, the present manager of the Matador division of the ranch, who very courteously bade me make myself at home around these interesting headquarters. I spent some thirty minutes attempting to take photographs here in the late afternoon before resuming my journey.

Driving back to the highway, I turned northeastward to the town of Matador little more than a mile away. This small town had grown up in the heart of the ranch and had become the county seat of the then newly organized Motley County, some sixty-three years ago. As is the case at Roaring Springs, there are a number of individually owned tracts of land at Matador, but over behind these possessions of the little men in nearly all directions are the almost limitless acres of the Matador Land and Cattle Company.

Before H. H. Campbell and his associates had gained control of this area in and near Motley County, a number of smaller cattlemen had herded their cattle here as was the case in the Spur Country. As we have already seen, Jim Hall had built up a considerable herd northwest of present Matador only to move his cattle to a point southwest of Spur and make them available for the organizers of the Spur Ranch. Wiley and Coggins had also ranched in west Motley County for a while, only to sell their Jingle-Bob cattle that came from the herds of John Chisum to the men who laid the ground work for the Matador Ranch. George Baker was with Hall northwest of Matador; W. B. Champlin was somewhere on the Pease River, and Houston Bros. were in Motley County—all of them by or before 1881. Perhaps there were others that have been forgotten, but undoubtedly the list of little cattlemen in Motley County would have been longer except for the fact that Campbell and associates began the formation of the big ranch so early.

My friend W. M. Graham, late of Matador, is one of the pioneers who made his first trip up here to the headwaters of the Pease River before any of the cattlemen had arrived. Graham came on a buffalo hunt in 1877, while the last great assault on the North American bison was in full swing. He returned to the West in a few years and has become a prosperous independent rancher in this land of the Matadors. Harry Campbell, the son of the man who originated the great company, also is an independent rancher still holding forth and doing well in this country that has grown up around Ballard Springs.

I spent the night in Matador and the next morning turned eastward down the highway toward Vernon. Some twenty miles northeast of present Matador was old Tepee City, the first post office in the Pease River Valley. It was established in 1879, when many of the first ranchers were still arriving on the newly deserted buffalo range. The urge to see what was left of this old trading post caused me to turn off the highway some eight miles to the left. The one time frontier center was on land that is now within the Matador pasture. So far as I could see, Tepee Creek, the flat sandy stream that during rainy seasons ran near by, had erased all trace of the old village. A well made stone marker that stood on the rise to the west of the stream seemed to be the only present day reminder of this early post office where cowboys from over more than 5,000 square miles of West Texas range once rode in to stuff their saddle pockets full of mail. George and Nora Cooper, who live in Matador, may be the only surviving persons who lived at Tepee City during its younger days. I had had an interesting interview with them before leaving Matador a few hours before looking up this old time trading post.

13—Ranch Lore from the Pease to the Wichita

After spending a while here at old TePee City, I turned back to the highway and continued my trek eastward. It was only a few miles along this paved thoroughfare to the broken country near the Tongue River. The bridge over the dry sandy bottom of that stream was almost on the east line of Motley County. Here, and covering some 79,000 acres[1] to the south, is the Tongue River Ranch of the Swensons. Early it was known as the Scab Eight Ranch, so called by the cowboys because the brand, 8=8, often scabbed over and peeled off between the equal marks. More than forty years ago when the Swensons purchased the property,[2] they changed the brand to their own S M S.

In Cottle County eastward of the Tongue River, my highway passed through a farm area where cotton production seemed to predominate, but nowhere in this part of Texas has the humdrum daily routine of farm life entirely dimmed the former lustre of the

1. W. G. Swenson supplied this information from records in the company offices at Stamford, Texas.
2. Actually not all of the ranch lands were purchased at one time. Much of it was purchased in 1902 as shown by the company records at Stamford, Texas.

old West. In more than one spot harvest hands in the fields may stand waist deep in cotton stalks and look across the fence where cowboys round up the fine white-faced cattle on one of the greatest ranches left on earth. The great Matador property, not far to the north of my highway, stretches a dozen miles down the Tongue River into this half modern and half ancient Texas county. Here, without a single cross fence, the Matadors have a pasture that covers 110 square miles.

The car moved easily along the concrete pavement that overlooked both plantations and ranch lands. Enough of the old range had gone under the plow to excite again one's curiosity as to the future status of beefsteak. Cottle County which was once exclusively grasslands now produces 15,000 to 35,000 bales of cotton per year and many tons of beefsteak in addition.

Shortly I came to the town of Paducah and continued eastward, still through farm lands. In a few miles the terrain, especially on the south of the road, became more broken and farm lands gave way to ranches.

To the northeast of Paducah and extending into Childress and Hardeman counties, was the early day ranch that ran the O X brand. During open range days, probably beginning in 1880, this brand was worn by ten to fifteen thousand head of cattle grazing an area around the junction of the north and south forks of the Pease River—an area half as large as a modern county.[3] George Merchant started the brand about 1875 in Denton County and after three or four years, sold it to Cairns and Forsythe Brothers[4] of Gainesville. The purchasers trailed their herds bearing principally this O X brand up to the forks of the Pease some ten miles north of my highway and turned them loose about the year 1880. Their headquarters, said to have been the first house built in Childress County, was northward of the junction of these two rivers. Several smaller herds were thrown with the Cairnes and

---

3. Gus L. Ford, **Texas Cattle Brands**, 87, 196.

4. **Ibid**, 196. Ford's **Texas Cattle Brands** credits A. J. Forsythe with full ownership of this old ranch but the **Stock Manual** (p. 265) by George B. Loving, published in 1881, mentions the owners as Cairns & Forsythe Bros. Loving's information was obtained direct from the various ranchers themselves. This is Loving's note about Cairns & Forsythe Bros.: "Post Office, Gainesville, Texas, Ranch Post Office, Tepee City, Motley County, Texas Ranch, Forks of Pease River, Horse brand OX left shoulder."

Another bit of information about Cairns & Forsythe is recorded in Vol. I, Page 7, "Marks and Brands" of Clay County, Texas. On October 7, 1878, they registered their H. F. (connected) brand in that old record book and gave Clay County as their residence. There seems to be no evidence that they ever moved this brand of cattle up to the Forks of the Pease River.

Forsythe cattle.[5] The most important of these was a herd bearing the M D brand and belonging to Swearingen Brothers.

Cairnes seems to have sold his interest in the ranch but the Forsythes held on until 1894 and sold to D. D. Swearingen and associates. Swearingen with one or more partners held forth with the old O X brand until 1930.[6] A sizeable remnant of the ranch is now the property of W. H. Portwood of Seymour, Texas. This Portwood Ranch, that runs well into the thousands of acres, is located north of my highway between Paducah and the village of Swearingen.

Another design that has burned itself into the early life story of the country eastward of Paducah is the brand known as "the Moons." As the O X brand has lived better in folk language than have the Forsythes who brought it west, so has "the Moons" stayed in the stories of the people better than have any of its owners.

Charlie Cannon from the village of Bolivar forty miles north of Fort Worth is correctly credited with having brought the brand to Cottle County. He was already ranching on the north Wichita River southeast of present Paducah, as early as 1879.[7] Cattle that bore the Moon brand also belonged to his former neighbor, W. A. Hughes who pastured his herds on Denton Creek a few miles west of Charlie's old home.[8] What relationship existed between them or what kind of a deal was made so that each might continue to run the same brand we may never know, but the Moon brand soon spread over many square miles of east Cottle County and it has stayed in the minds of local folks almost as well as the Matador's V. About 1883, Cannon lost his life in an incident at Seymour.[9] Some time following this unfortunate event his cattle were sold to Lindsey, Bedford, and Hinton, and the brand stayed on the cedar covered hills of Cottle County. Some 10,000 cows[10] are said to have spread over a range perhaps as large as the Pitchfork Ranch of today. They were sold to an Englishman by the name of Maud, who sold them again in the middle 1890's.

Like many of the ranches that began on open range the size of the Moon-branded herds began to show a decrease when land

5. George B. Loving, **The Stock Manual**, 235, 269, 271. Swearingen Bros., Sherwood & Hinman and F. R. Sherwood all shared the range with Cairns & Forsythe Bros.
6. Gus L. Ford, **Texas Cattle Brands**, 196.
7. **Fort Griffin Echo**, December 13, 1879.
8. George B. Loving, **The Stock Manual**, 224.
9. **Fort Griffin Echo**, July 14, 1883.
10. A note giving some of the history of the moon brand is found in Gus L. Ford's **Texas Cattle Brands**, 117.

ownership and wire fences prescribed the exact limits of a ranch. However, the 60,000 acre remnant of the Moon Ranch that passed to W. Q. Richards in a few years was no midget in itself.

Richards perhaps will be better remembered than many of the other men who owned the Moon brand. He was a more colorful Westerner than most of the men of his day. Folklore always builds interesting tales about such a man. Once some difficulty separated him from his wife. According to the current story,[11] he agreed to a separation but seemed to blame himself for the trouble. In true western fashion he built a fence across the middle of the ranch, gave his wife the half that included the old headquarters and settled down to ranching on the other half. The former husband and wife lived as good neighbors across the fence from each other throughout their remaining tenure of the property. Richards sold out to W. O'Brien in 1920 and even until this day, O'Brien runs the Moon brand on a ranch in New Mexico.

But Richard's cattle dealings were not confined to this famous brand alone. South and east of the "Moons" George Brandt of Kansas City and others built up the 3D Ranch—more properly called the Hesperian Land and Cattle Company. As a kind of beginning, the F. P. Knott surveys covering 40,000 acres in southeast Cottle County were purchased by Brandt for the very small price of twenty-five cents per acre.[12] Next, the Mary Ann Cook survey and other lands to the west were added, and then a rather large slice of east Cottle County real estate brought the total holdings of the 3D Ranch within striking distance of 100,000 acres. The ranch lay in the shape of a giant L fitting snugly into southeast Cottle County and even extending over the edges a little. It fitted against the Moon Ranch both to the south and east and carried on the usual neighborly relations from its beginning about 1880 until near the turn of the century.

Soon after Richards had bought the Moons, he began, figuratively speaking, to look over his south and east fences—and to do a little thinking. He is said to have been one of those lightning-like traders that deal as quickly as a Westerner can draw his gun.

According to a current story—which may be more folklore—he made an unannounced visit to some unknown point—perhaps to see his banker—and returned with as much secrecy as he had gone. In a day or two he had an opportunity to see the owners of the 3D Ranch. They talked casually of selling and inadvertantly

---

11. Interview with R. Ernest Lee.
12. Interview with John Gibson.

made a proposition. With as much suddenness as some elements of modern warfare, W. Q. Richards became owner of part of the 3D land and all of the cattle "as quick as your hat could have hit the ground." He soon sold part of his newly purchased cattle to great advantage and thereby knitted back the raveled edges of his much expanded fortune.

The story may be true. Often men of the old West shot from the hip and traded the same way. But regardless of the accuracy or inaccuracy of detail, Richards did become the owner of the 3D cattle and part of the 3D lands only two or three years after he had purchased the Moons.[13]

The full details of how Richards sold all of his ranch lands are not available, but much of the old Moon and 3D ranches a little later became the property of Tom Burnett, son of the now famous Burk Burnett who rose to the magnitude of a financial giant as the builder of the 6666 Ranch. Both Burk Burnett and his son Tom have passed away but their ranches are still an important part of the old West.

The Tom Burnett properties are known as the Triangle Ranch from their brand (the triangle).

Four miles along the highway east of Paducah I came abreast of this Triangle Ranch on land where former ranchers had once branded the "Moons" and the 3D. The present ranch joins the highway on the south for some fifteen miles or more. The Triangle Ranch is about half as large as the 6666—but let us discuss the magnitude of these several ranches in a later chapter.

In the east part of Cottle County, principally north of the present highway, was one of the several ranches of the early 1880's in which the operator did not own the land. Here, joining the old 3D on the north, Bob Wright and his brother William Crow fenced a tract of land seven miles wide east and west and thirteen miles long, without owning an acre—according to old cowboys who worked on neighboring ranches.[14] William Crow Wright had some 1200 head of cattle (in 1881) branded with what was called the Chain Seven.[15] It consisted of a long straight line burned on a cow's side from shoulder to hip with a figure seven burned over the line at each end and another over the middle. Bob Wright with perhaps as many cattle as his brothers, grazed the same area as these Chain Seven cattle. Bob's brand was known as the

13. Local interviews in Cottle and Foard counties with John Gibson, Lee Ribble and others.
14. Interview with Jim Olds of Crowell, Texas.
15. George B. Loving, *The Stock Manual*, 26.

Chain Half Circle. It had the long line just the same as did the brand of William, but instead of the sevens, Bob branded with half circles over each end of the line with no circle in the middle.[16] Shortly, we shall relate a rather odd incident that involves this old brand of Bob Wright. Meanwhile let us resume our journey along the highway east of Paducah.

A car seems to glide with ease along the highway through this ranching area perhaps because the slight traffic makes it easy for the driver to relax. But the endless succession of hills perpetually green with cedars also contributes to the driver's quiet comfort.

In a short while, I entered Crowell some thirty five miles from Paducah. Here I talked to Jim Olds, an early day cowboy on the old R 2 Ranch. The old R 2's were the property of Stevens and Worsham as early as 1879.[17] Their immense range covered most of the country northeast of Crowell all the way to Red River.[18] Jim had worked as outside man for this ranch in the middle 80's and had learned to know most of the ranchers and many of the cowboys across King, Cottle, Foard and other nearby counties.

But one of the unusual incidents of Jim's life as a cowboy had little to do with the R2 Ranch. An immense outlaw steer that had belonged to the Witherspoon Ranch had run wild over the countryside for several years.[19] He was originally one of Bob Wright's chain half circle steers that was born in the Pease River Breaks along the north end of Wright Brothers' old pasture some twenty miles northwest of present day Crowell. The Wrights had sold their cattle to Witherspoon some time in the early 1890's. This particular steer refused to be driven in the herd when the Witherspoons made the purchase, and in fact refused to be penned or to be driven in any herd, and he developed a reckless disregard of wire fences. The old steer weighed something near 1600 pounds, and it would have been rather risky for any cowboy to have thrown a rope over his head. The cowboys for some distance around knew about the unruly creature and some of them had learned to be a little careful not to rush him into any close places.

With such a reputation, the steer ran at large until he was about sixteen years old and had become not very different from a wild beast. It was about 1905 when a price was finally placed on

16. Interview with Jim Olds.
17. Interview with the late George Cunningham of Henrietta, Texas, who worked for the R2 Ranch in 1879.
18. The Jim Olds Interview.
19. The story of the outlaw steer came from Jim Olds.

his head and efforts were made to capture him. Henry Ferguson, one of Witherspoon's cowboys, found him one day in the Pease River Breaks northwest of Crowell. He was ready to fight or run whichever the occasion demanded.

The date 1905 was long after the days of the open range and Stevens and Worsham had been forced to retire from their great range to the northeast of Crowell. They had purchased a small pasture to the northwest of that town—some of the very land that Wright Brothers had fenced in early days. The outlaw steer that grazed in the broken country near the Pease River was in fact born on land that the R 2 Ranch had thus acquired. The steer intended to remain on his native grazing area regardless of ownership. Henry Ferguson had different ideas.

Jim Olds had become local foreman for this R2 Ranch where the Chain Half Circle steer ran wild, and Henry consulted him about the capture of the old outlaw. Jim lent a couple of cowboys to Henry and wished him good luck in his undertaking.

The steer sighted his pursuers and headed for the worst of the brakes. Henry knew the danger involved in throwing a rope over the head of such a desperate beast and decided against such procedure. Gun powder was the only answer to the problem and both Henry and his two assistants had the answer ready. A quick shot wounded the 1600 pounds of rebellious beef and rawhide, and the desperate animal may have had for a fleeting moment an inclination to turn and fight. If so, he was too slow to make his decision effective. It was no time for weak-kneed decisions on the part of the cowboys or for sentimental applications of some fanciful code of sportsmanship. The mad steer most likely meant to use his horns.

Such a dramatic climax was spared by the united action of the three cowboys. Quick and judicious use of lead and gun powder crumpled the steer to the ground. The old outlaw that had lived to himself and had borne Bob Wright's Chain Half Circle brand for sixteen years had come to the end of the trail.

Henry, Jim and his cowboys skinned the self willed old creature and carefully saved his horns. Some years since the occurrence, a boy of artistic talents has painted a life size cardboard dummy of the massive steer mounted with the actual horns that Henry and Jim had saved. This life-like reproduction of the outlaw was for a long time kept on the balcony in Ferguson's Drug Store here at Crowell. Henry and Jim, now many years after their rip-roaring days as cowboys are over, have posed with the dummy steer in a picture that is reproduced elsewhere in this volume. Probably there were not enough ropes in the county to have held the steer still enough for such a picture had he been captured alive.

By an odd chain of association the old steer ties together some of the important ranch history of the country that lies around Crowell. The rugged old outlaw was killed in the Pease River Breaks northwest of Crowell, but his rightful home was the Witherspoon Ranch south and southeast of town. The Witherspoon property was often known as the "9" Ranch because of its principal brand—three nines with the middle nine lying down. But nobody was ever able to burn this nine brand on the big steer's hide since he so successfully eluded capture.

The Witherspoon Ranch was also known as the ULA Ranch, a fact that takes us back another step in the story of Foard County ranches. This ULA brand was originated by Tom Atkinson of Jack County, who began the brand soon after his baby daughter Ula was born in the year 1870.[20] Curtis Brothers and Leonard bought a portion of Atkinson's cattle interests in 1880 for $40,000.[21] The purchase included only the Atkinson interests on Good Creek, which was in the area that is now Foard County southwest of Crowell. The ULA brand went along with the deal. It was not the chief brand of the Curtis partnership; they emphasized mostly their Diamond Tail brand. Their principal ranch was on Buck Creek across Red River north of present day Childress, and it was on that range that they soon concentrated their interests. The ULA brand was disposed of, becoming Witherspoon property and remaining in Foard County.[22] But these new owners of the ULA brand who also owned the Chain Half Circle outlaw steer spread their range over most of the south half of Foard County and stayed long enough, like some of the other ranches, to become a part of the folklore of the country. Much of the ranch is farmland now except that a sizeable portion of it is one of the several ranches now belonging to Ferd Halsell.

Another large ranch near this Witherspoon range had for some time attracted my attention. I had heard too much about

20. Interview with Mrs. Ula Carver, Wichita Falls, Texas.

21. **Fort Griffin Echo**, November 27, 1880. Mrs. Carver, the daughter of Tom Atkinson does not now remember that the ULA cattle were ever sold to Curtis and associates but a news item in the old **Fort Griffin Echo** of the above date reports the sale and states the price as $40,000. Also the old **Stock Manual** (p. 56) published about six months later shows that the ULA brand at that time belonged to **Curtis Bros. & Leonard**.

22. The Jim Olds interview.

the fine Aberdeen-Angus ranch of the late Tom **Masterson** over in King County southwest of Crowell not pay it a visit. **Masterson** did not run the whoop-'em up, run-at-breakneck-speed kind of a ranch.He believed in being quiet with cattle—especially with the Angus variety—and his great success offers strong testimony to support his ideas. A cowboy who wanted to ride with a whirling loop and to yell and excite the cattle could not work for **Tom Masterson.**

The old-time round-up with its fast beating of horses hoofs has no place on this Masterson (J. Y.) Ranch. Here the old-time round-up has been slowed down to a walk. Men sit or move quietly on horseback around a herd of cattle, and the required number to be cut from the herd is moved easily in the proper direction with as little excitement as one finds around the milking pen of dairy cows.

Masterson had cut his ranch into a number of small pastures and had some little land cultivated in feed crops. Many trench silos had been made and the methods of progressive ranching were followed generally. A neighbor who had certain prejudices against the use of silage on a ranch one day asked the owner of the J. Y.'s if that kind of feed did not injure some of the calves. "The only calf that silage hurts is the calf that fails to get any," was Masterson's pointed reply.

Do these methods of Tom Masterson and the other owners of the associated ranches pay or are they just wild theory? Probably the only answer required is to point out the bare fact that the Mastersons in the Panhandle and in King County have built up interests in cattle much larger than the nationally famous Pitchfork Ranch. The ranch in King and Knox counties is usually stocked with some 6000 Polled Angus cattle and is said to be the greatest Polled Angus ranch in the world.

A visit to the Masterson Ranch required a side trip from my selected route but it was well worth the journey—and beside the real values involved, I had a sense of curiosity to see the ranch. Also I was invited to eat chuck well out in the middle of the pasture at Joe Meador's construction camp. Joe was contracting the building of some water tanks for the ranch and not a small feature of the camp was Humpy Briggs' made-to-the-measure-of-your-appetite cooking. Humpy was an artist with sour dough biscuits and there was not a tough bite of meat nor a bad frijole in the pot to say nothing of knick-knacks to polish off the finer touches of your appetite. But we run ahead of the story.

I drove the seventeen miles southward from Crowell to Truscott where the road turns off some three or four miles west to the Masterson Ranch. Once inside the first cattle guard, I glided along through hundreds of evergreen acres—green with cedars always, and of course green with grass according to season. There was much broken land to furnish shelter to cattle in the worst of winter weather. Several miles within the ranch were the excellent ranch home and headquarters. Then next there were the old Rock Corrals that were once the old "P 8" headquarters,[23] quite a landmark to old timers. Not very far ahead, I came to the middle Wichita River. Here was an actual river running some twenty or thirty feet wide with clear spring water that looked good enough to drink—but it wasn't. The water carried enough gypsum and salt to cause one to let his thirst wear on for awhile.

I traveled down to the river on almost the identical road that the old first mail line from Fort Griffin to Mobeetie followed sixty eight years ago, and coming into my road from the left were the tracks of a little known branch of the old MacKenzie trails. The old ruts were etched into the gyp rock formation so plainly that I could have followed the road for some distance across the pasture without difficulty.

It was two or three miles on to Joe Meador's camp where Humpy Briggs, the comedy-loving cook, soon yelled, "Come and get it." I picked up one of the tin plates and filled it from the different pots until the plate really needed sideboards. There were tin cups for coffee or water, and apricot pie for dessert—to say nothing of cookies and other delicacies. I sat down on a box with my plate in my lap and stuffed myself beyond all reason. After the meal there were a few camp yarns and then we walked down a short distance to inspect a large dirt water tank that Joe was digging.

He was building a dam across Farrar's Creek a few hundred yards below the mouth of Polecat Creek. Both streams had acquired their names in a perfectly natural way. George Brandt of the 3Ds, not long after 1880, made down his bed roll among the ranch hands who happened to be camped not far up the little creek above us. He awoke with the unpleasant feeling that something was chewing on him—and he was not mistaken. A "polecat" had bitten him in the head and he had the disagreeable prospect of contacting hydrophobia! The cowboys laughed at George and his "polecat" but it hardly seemed like good comedy to George.

23. Interview with Lee Ribble of Vernon, Texas.

He soon summoned medical aid and found that the little wild creature was infected, confirming his worst fears. At some expense and trouble he prevented the tragic results that might have befallen a less cautious person. Since the incident, cowboys , geographers, and all have called the little stream above us "Polecat Creek."[24]

Farrar's Creek, the principal stream on which Joe was digging the tank, took its name from John Farrar who began running his J F branded cattle on this area soon after the last herds of buffalo were exterminated.[25] He had lived at Ennis, Texas, and had kept his cattle on the open range of Baylor County for a while before venturing up here into King County.

The old J. F. headquarters were two miles down Farrar's Creek from the site of Joe Meador's camp. Farrar had built the rock corrals and a rock house at that place[26] because some buffalo hunters had left a few slight improvements for him to begin with. His camp site had been covered with great stacks of buffalo hides during the latter days of the great hunt.

John Farrar prospered in the cattle business on this King County ranch. He branded some 3300 calves here in 1888 and sold out to Mabry, Crawford, and Glasgow of Graham for $50.00 per cow. The calves went in the trade without extra cost. Soon Sam Glasgow sold his interest to the other partners.

The Mastersons came into possession of the land in 1898. Previously, in the 1880's, R. B. Masterson had ranched in old Greer County which is now a part of southwestern Oklahoma and had moved his cattle to Mobeetie in the east Panhandle. Then, just before 1900, he moved down into King County and bought something near 100,000 acres of this range that included Farrar's Creek and no little of the middle Wichita.[27] Here the partnership, R. B. Masterson and Sons, was formed and the family began to grow one of the larger cattle fortunes. In 1911 more than 100,000 acres of the old L X Ranch in Potter and Moore counties was purchased by the partnership, and by 1917 some 15,000 head of cattle grazed their excellent ranges. The property has been divided in later years, but the holdings have been kept intact and effectively still constitute a large ranch.

The late Tom Masterson for some time had charge of the King County ranch as already noted. As I drove the sixteen miles

24. The Lee Ribble interview.
25. George B. Loving, the Stock Manual, 224.
26. The Lee Ribble interview.
27. Gus L. Ford, Texas Cattle Brands, 138, 214.

from the first cattle guard to Joe Meador's camp, I could feel that the spirit of Tom Masterson was still with the ranch. The slick, fat black cattle grazed quietly by the roadside; not a cowboy was in sight, unless perchance, as a general utility ranch hand he drove toward some particular task down the road beside us.

Backtracking from the Masterson Ranch to Crowell, I came again to my highway, ready to continue the journey eastward. In making this side trip to the little kingdom of the late Tom Masterson, I had passed through the former domain of T. S. Witherspoon. As we have seen previously, this early rancher, owner of the above mentioned outlaw steer, had spread his ULA and Nine brands over a large area south and southeast of present-day Crowell. In time he acquired title to much of the land and settled down to a more stable form of ranching. But as I drove eastward from Crowell, I noted that many thousands of acres of this country to my right that once produced Witherspoon beef had gone under the plow.

On my left was another of the old ranches of the early West. It was the ranch with which Jim Olds had spent a number of his best years. Far to the north of my road across the Pease River and miles beyond the range of eyesight, was once this R2 Ranch of the big cattlemen, Stevens and Worsham. Of course, the present highway was not the exact boundary between the old Worsham and Witherspoon ranches. The range of the Worshams reached as far north as Red River and south all the way into this Crowell country. It extended eastward to Vernon and westward far beyond present Chillicothe—in fact the old headquarters were located on Groesbeck Creek a mile south of the site of Chillicothe.[28]

The town derived its name directly from one of Stevens and Worsham's hired hands. One of their cooks whose name was Jones had come to the wide open West from Chillicothe, Missouri. He settled down to cooking for the R2s for a while, but after a time decided to go in business for himself. Accordingly, Jones built a little store a mile down Groesbeck Creek from the ranch headquarters and named it Chillicothe after his old home town in Missouri.[29] The name stuck and the little town prospered and was lucky enough to become a railway station when the Fort Worth and Denver railroad built west.

Jim Olds who, as noted above, was one of the ace cowboys for

28. The George Cunningham interview.
29. The Jim Olds interview.

Stevens and Worsham at the time when Jones was one of the cooks was proud of the magnitude of the old R2s. "We branded 10,000 calves a year," he said, and his voice carried with it a pride akin to that which goes with ownership.

It was a great ranch and its beefsteak once helped to supply the needs of the American market; but the little store built by Jones the cook, the railroad, and the nesters all grew together until there was scant room left for the Stevens and Worsham cows. A small remnant of the great herds was moved to the ranch purchased northwest of Crowell—the ranch where the outlaw steer was killed—but most of the cattle had to be disposed of in other ways. And now the country around Jones' store, like much of the land once grazed by the Witherspoon cattle, has turned to the production of acres of cotton and wheat and row on row of the various sorghums. Science has helped here to make the victory of the farm complete, for it is at the very town of Chillicothe, once the center of the great ranch, that A. B. Conner and Roy Quinby have made such remarkable developments in the production of non-saccharine sorghums adaptable to semi-arid West Texas.

Soon I drove into Vernon beyond the borders of the old Withspoon and Worsham ranges. Here in the valley of the Pease River are some of the most fertile lands to be found on the prairies of the West. Long ago these rich acres were placed under the plow, although they were first part of the immense West Texas open range.

Dan Waggoner was already on the lower Pease by 1881[30] and in northwest Wichita County as early as 1879.[31] He had come up from Wise County to the country around Wichita Falls in 1871[32] and had moved up to this Pease River area some seven or eight years later. Shortly after he made this latter move, he began to graze perhaps a half million acres across Red River chiefly in the west end of present day Tillman County, Oklahoma. The range was held under a grass lease covering Indian lands of the Kiowa-Comanche Reservation. It was put under fence and held well past the year 1900. Waggoner was one of three ranchers whose cattle ranged in Tillman, Cotton, and adjoining counties as they are known on the map of present Oklahoma. Waggoner held the west end of this large grazing area; Burnett joined him

30. George B. Loving, **The Stock Manual**, 52.
31. Interview with the late W. D. (Shinnery) McElroy of Harrold, Texas.
32. The McElroy interview.

on the east, and E. C. Sugg held forth to the east of Burnett. The combined grass leases of the three ranches extended from the north fork of Red River northeast of Vernon eastward to the 98th meridian and from a line a little south of the Wichita Mountains all the way down to Red River.[33]

The three ranchers learned how to maintain friendly relations with the Kiowas and Comanches. Waggoner's Three D's brand, Burnett's Four Sixes, and Sugg's O H Triangle were pretty generally respected among the Indians.

Ranching on these Indian reservations was probably no more hazardous than ranching elsewhere and the chores of cattle raising settled down to the common routine. Quanah Parker, the former Indian chief who had played, as it were, a game of hide-and-seek with Mackenzie on the plains, was employed here by Waggoner in a kind of extra ranch hand capacity.[34] Often the old warrior could be seen riding the Waggoner ranch line in his old buck-board—always accompanied by his youngest and favorite wife "Too-nicie."

The north and east sides of the Waggoner pasture were fenced soon, making it necessary for men to ride along them on regular tours of inspection. Charlie Taylor once rode these fences for the Waggoners, spending many lonely hours with nothing more entertaining than common-place barbed wire and fence posts. He lived in a dugout that was located on Waggoner's east fence some six or eight miles northwest of present day Grandfield, Oklahoma. George Newton, fence rider for Burnett, shared the dug-out with Charlie. George rode Burnett's north and west fences. The latter was the same as Waggoner's east fence and was a common assignment to the two cowboys.

This long lonely barricade of barbed wire began at Red River, passed some five or six miles west of the site of Grandfield, up past the dugout shared by the two ranch hands, and many miles beyond. It came to an end more than thirty miles north of Red River, terminating at the northeast corner post of Waggoner's great pasture. The same corner was Burnett's northwest corner post.

George and Charlie set up the bony remains of an old steer head on this common corner post to be used as a silent means of communicating between themselves. It was understood that the

33. Interviews with Charlie Taylor and Bob McFall both of Wichita County, Texas.
34. Interview with the late W. H. Portwood of Seymour, Texas.

last one of them who came to the steer head should point the nose of the animal in the direction that the cowboy's duties had taken him. From this slight signal the other cowboy as he came riding up to the corner post could know instantly which way his buddy had gone. It was some slight comfort to the two old cowhands, who often spent several days up the fence lines away from each other and sometimes perhaps weeks in which they saw no one else.[35]

Dan Waggoner was well aware of the uncertain nature of the grass lease on Indian lands and had much earlier than 1900 made ready for the day when the authorities at Washington would force him to provide other pasturage for his cattle.

For a long time he had shared the common feeling of cattlemen that the range should be left open. In about 1884 while the Ikards and Harrolds were moving their cattle across Waggoner's ranch northwest toward old Greer County, the several big ranchers one day ate at the same chuck wagon. The Ikards and Harrolds explained how the "little men" and nesters had crowded in on them making their move to a new range necessary. Waggoner agreed that the cattle business had come to a sad state. He said, "A

35. The story was told me by Charlie Taylor himself.

man can't afford to even lease the land, much less to buy it."[36] But he was feeling the pressure of other cattlemen and evidently soon saw that the only possible defense was to buy land. Over a period of years he purchased a large body of land in the south part of Wilbarger and the north part of Baylor counties. His immense holdings here of 510,000 acres stand as one of today's greatest ranches.

I turned southward from Vernon along a paved highway into the heart of this great ranch. Two beautiful ranch houses lay to the right of the road—one the former home of Electra Waggoner-Wharton, the other the home of E. Paul Waggoner. Death has claimed the elder Waggoners, Dan and W. T., but the mammoth estate is still held together under a single management.

Some of the Waggoner Ranch story is told in pictures elsewhere in this volume, but this narrative must move on to some other colorful events which await us.

---

36. Interview with the late Lewis Richardson of Jacksboro, Texas, who was one of Harrold's cowboys.

14—Burk Burnett Did Not Own Boom Town

I came back out of this great cattle empire to **Vernon** ready to continue my journey through the cow country. It was some twenty miles eastward from Vernon to the town of **Electra**, named for W. T. Waggoner's only daughter. Eight miles farther travel along the highway, placed me in front of the ranch home of the late Tom Burnett. I went in the big gate to the ranch house. Not a human soul was there, as my efforts to arouse someone—anyone—soon convinced me. Even my attempt to photograph the interesting dwelling house did not turn out well and had to be done over at a later date.

But just the visit to the home of this unusual cattleman and character of the old West was provocative of thought. He had built this modern—if not elaborate—structure on the same spot where his father had placed a little headquarters building seventy-five years ago and near where the first dugout was scooped out of the ground two years earlier than that. Tom was only four years old when the Burnetts first broke dirt in this area.

Burk, his father, began ranching here in 1875[1] and spread his 6666 brand over the hills of Wichita County for a full quarter century afterwards. Soon after 1880, the elder Burnett had secured a grass lease across Red River in Indian Territory—now Oklahoma—which at one time amounted to nearly 300,000 acres.[2] This new addition to the domain of the Four Sixes caused the center of the range to move northward and gave rise to a change in the location of the Burnett headquarters. Accordingly, a ranch house was built only a mile south of Red River and some four miles northwest of the location of the future oil town of Burkburnett, mentioned earlier in this account. It was while this new house was the nerve center of the Burnett cattle interests, that the 6666 brand was to see one of its greatest periods of expansion. At least, 10,000 head of Burk Burnett's cattle grazed on the leased Indian lands north of Red River[3]—and there were other thousands on the Texas side of the river.

During this eventful time young Tom Burnett became old enough to make a ranch hand and came west to engage in the cattle business as an employee of his father. It must have been one of the top ranking adventures of Tom's life to become a cowboy in this new wild frontier country,[4] for he was only sixteen years old. To make his experiences even more colorful he was located north of Red River where he was in almost daily contact with the Indians from the nearby government reservations. He learned to know dozens of these red men personally and soon learned to speak the language of the Comanches. He came to know Indians and Indian customs perhaps as well as any white man of his day.

At the age of twenty-one Tom became wagon boss, and in spite of his extreme youth the cowboys respected him as a real leader and a capable cowman. It was not long after he had become wagon boss that he married Miss Ollie Lake of Fort Worth. His bride came with him to the cow country to live. The young couple made their home at the Burnett ranch house on the south

1. John M. Hendrix, "Tom Burnett," **The Cattleman**, May, 1939, 16. Burk Burnett's 6666 brand was registered in Clay County ("Marks and Brands," Vol 1, 7) July 22, 1876. Since Wichita County was not organized until 1882, all cattle brands within its confines prior to that date were registered in Clay County. Registry of a cattle brand lagged somewhat behind the date when a cattleman entered a new county. Hence Burk Burnett's brand book entry in 1876 is entirely in harmony with the statement in the Hendrix article that the now famous 6666 ranch began in Wichita County in 1875.

2. Interview with the late Bob McFall, a former Burk Burnett cowboy.

3. John M. Hendrix as cited above page 18.

4. The story of Tom Burnett is told in the article by John M. Hendrix as cited above.

side of Red River where the daily routine was constantly en-
livened by a succession of cowboys, Comanche Indians and
wayfaring frontiersmen.

After living at this ranch home for some ten years, there was
born to Tom and Ollie a daughter that was to become one of the
important personalities of the cattle world. The little girl, Anne
Valliant Burnett, was one day to become the most important person in the ownership of two great ranches—but we are running
ahead of the story. It was about ten years after Tom and Ollie
moved to the ranch house on Red River that the Burnetts were
brought face to face with one of the most difficult problems of
their ranching careers. The lands to the north of Red River were
to be thrown open to settlement and, along with this change in
governmental policy, the grass leases that had been a rich source
of revenue to the Four Six empire were to be discontinued.

As told earlier in this story, Burk Burnett did not meet this
crisis lying down. He went to Washington in 1901 with Joseph
Weldon Bailey, his attorney, and laid the matter before President
Roosevelt.[5] A modification of the order came about — probably
as a result of this conference—which permitted Burnett, Waggoner and perhaps others to retain a part of their holdings. As a
result the ranchers were enabled to work out their problem of
finding range for their extensive herds of cattle.

Anticipating the difficult situation, the 8 Ranch in King
County had been purchased a year before the meeting with the
President; and about the same time the 106,000 acre Dixon Creek
pasture in Carson County was added to the Burnett's range to
make the position of the 6666 brand secure in the cattle world.[6]
The day of open free range had gone and extensive grass leases
were about to be a thing of the past. The time had come when a
man who wished to stay in the cattle business must own land—
and plenty of it. Burk Burnett's two large purchases, with some
small later additions, amounted to a third of a million acres. He
had chosen to stay in the business of producing beefsteak on the
hoof—and he had not been timid in carrying out his decision.

However, as a part of this story of Burk Burnett, the closing
out of the grass leases north of Red River was by no means a humdrum affair. It involved one of the most colorful wolf hunts ever
staged in the Southwest. Theodore Roosevelt—then President
of the United States—was entertained in the heart of these leased

5. C. L. Douglas, **Cattle Kings of Texas**, 356-7.
6. Interview with John Gibson of Paducah, Texas.

THE CATTLEMAN'S

> *Home on the Range*
>
> *is no longer a dugout.*
>
> *Often not ranch style,*
>
> *his house stands up like*
>
> *an ancient castle.*
>
> *The seventy-five-year-old*
>
> *Waggoner Mansion at Decatur, Texas,*
>
> *and the somewhat more recent*
>
> *Four Six Ranch House*
>
> *at the heart of 200,000 acres*
>
> *of King County grasslands are*
>
> *magnificent examples.*

PITCHFORK RANCH HOUSE—85 miles east of Lubbock, Texas on Highway 82. See chapter 11.

Photo by Paul J. Pond

OLD WAGGONER MANSION
at Decatur, Texas
For Waggoner Ranch Story
See Chapters 1, 13 and 14

Photo by Paul J. Pond

Tom Burnett Ranch House at site of original Burnett headquarters established 80 years ago—18 miles northwest of Wichita Falls, Texas, on Highway 287. See Chapter 14.

Photo by Paul J. Pond

The Four Six Ranch House at Guthrie, Texas—95 miles east of Lubbock. Burk Burnett intended this (in 1917) as the finest ranch house in West Texas. Story told in Chapter 3.

Zacaweista, first of the fine homes on the Waggoner Ranch. It was the home of Electra Waggoner Whorton and now the home of her son, A. B. Whorton, Jr.

Photo by Dick McCarty

Santa Rosa, the magnificent home built by E. Paul Waggoner on the 500,000 acre Waggoner Ranch south of Vernon. It is now the home of Mr. and Mrs. John Biggs. Mr. Biggs is manager of the ranch. The Waggoner Ranch story told in Chapters 1, 13, and 14.

Photo by Dick McCarty

Ballard the buffalo hunter camped here in the 1870's. The Matador Ranch headquarters were established here in 1879. This house is of more recent date. The place is now the property of the Rock Island Oil and Refining Company. It is located near Matador, 60 miles east of Plainview, Texas, near Highway 70. See Chapter 12.

Photo by Paul J. Pond

CATTLEMAN AND COWBOY

share the great variety

of dwelling places that loll at ease

on today's immense empires of grass.

Some of these dwelling places

speak of home;

some speak of room and board;

some speak of the history of yesteryear;

some tell a plain tale of hammer and saw.

In the area once known as Little Arizona, the Ed Bateman Ranch House is located 100 miles southwest of Wichita Falls, Texas. For stories of Little Arizona see Chapter 1.

Photo by Paul J. Pond

The Masterson Ranch House in northeastern King County about 100 miles west of Wichita Falls. The ranch covers 100,000 acres in King and Knox Counties. See Chapter 13.

Old Charles Goodnight Home at Goodnight, Texas, 40 miles southeast of Amarillo. Here Col. Goodnight spent many of his later years. The house with some 10,000 acres of ranch land is now a part of the Mattie Hedgcoke Estate. See Chapter 15.

Photo by Paul J. Pond

The old Henry Clay (Hank) Smith native rock house with its 22-inch thick walls in Blanco Canyon 10 miles north of Crosbyton, Texas, or about 50 miles northeast of Lubbock. In 1877 it was the last house west between old Fort Griffin and Fort Sumner, New Mexico. The Hank Smith Ranch was never larger than 5,000 acres but the ranch house is important because of its historic value and its unique owners. See Chapter 8.

The "8" Barn 10 miles east of Guthrie on the 6666 Ranch. This barn has been remodeled since it was a part of the old "8" headquarters in the 1890's, but it is still a landmark of early ranching in King County. The great cutting horse, Old Hub, spent his later days here. See Chapter 2.

The little Bunk House at the "8" Camp on the 6666 Ranch in King County 10 miles east of Guthrie on Highway 82. This plain little house, once a part of the old Q. B. headquarters, was moved to its present location in the 1890's. See Chapter 3.

The Jack Parkey Ranch House, some 20 miles southwest of Wichita Falls, Texas. The Parkey ranch lands were once grazed by the herds belonging to Harrold Bros. and to W. S. Ikard more than 75 years ago. See Chapter 1.

The Ross Ranch House 95 miles southwest of Wichita Falls, Texas, on Highway 82. This colorful area was once known as Little Arizona. See Chapter 1.

The Pitchfork Headquarters as it appeared 25 years ago.

The Pitchfork Bunkhouse at the headquarters 85 miles east of Lubbock. Individual rooms are provided for the cowboys like a hotel or boarding house. See Chapter 11.

Historic little house on the Swenson Ranch about 10 miles north-west of Spur, Texas. It was part of the Spur headquarters before the Swensons became owners. See Chapter 9 for some of the Spur Ranch story.

The old spring house on the Swenson Ranch near Spur, Texas. It was once a part of the old Spur Ranch headquarters.

Monument to the memory of the princely cattleman, Boley Brown. It is out on the range where he died—35 miles east of Post City, Texas. Above is the old home of Boley Brown. For the story of this great cattleman and cowboy, see Chapter 5 and the concluding pages of the book.

The long-bearded Scotchman, O. J. Weirn, founded the Two Circle Bar Ranch, later changed to the O-Bar-O. The present little ranch house 45 miles east of Post City, Texas, is the administrative center from which the heirs of Bert Wallace now operate this former ranch empire of O. J. Weirn. See Chapter 4.

The Morgan Jones Ranch House on Highway 380 about 35 miles east of Post City, Texas. See Chapter 5.

The Yellow Houses—southwest of Littlefield. The most noted camping place on the South Plains. In succession—a buffalo hunter's camp, XIT Headquarters, a Littlefield Headquarters, and now the George White Ranch Home. See Chapter 7.

The Cowan Ranch House some 35 miles southwest of Wichita Falls on Highway 82. See Chapter 1.

COWBOYS AND COWHORSES

are the soul of·both
work and play in the
Big Ranch Country.

Old Hub the great
cutting horse thrilled
the West in his day.

Wild horses, good horses,
common horses—tell a
varied story in this land of
romance and beefsteak.

LAST OF WILD STALLIONS captured about 25 years ago by 6666 Ranch Cowboy, Boog Graves, in the Brazos River Breaks of Northwest Stonewall County. The horse named "Nester Boy", made a good cowhorse on the 6666 Ranch. The photographs merged here were taken several years apart.

PALS IN THE OLD WEST OF SEVENTY YEARS AGO—Old Hub and Sam Graves. When Hub was more than 20 years old, as shown in this picture, he dazzled the cow country with his brilliant victory at the cowboy reunion at Haskell, Texas. The rider is Sam Graves. See chapter 2.

George Humphreys on Hollywood Gold. The horse was a gift from Tom Burnett in 1941. Mr. Humphreys is manager of the 6666 Ranch in King County, a position which he has held for more than 20 years. Hollywood Gold has sired many of the finest cowhorses in the West. For the 6666 Ranch in King County see Chapters 2 and 3.

Jack Spencer and his favorite horse, Sailor (at left). Jack was wagon boss for the Four Sixes in King County. He has won some enviable prizes with Sailor. See Chapter 3.

Bud Arnett on Red Bird. Mr. Arnett was manager of the "8" and "6666" ranches for nearly half a century. For many years old Hub was his favorite horse. See Chapters 2 and 3.

A MUCH LOVED ENGLISHMAN who knew the cow business — Billy McClaren, manager of the old Q. B. Ranch which was sold to the "8" Ranch in the 1890's. McClaren later served as wagon boss for the Four Sixes in King County. Note the long coat and special made boots, characteristic of his wearing apparel. The Q. B., "8" and Four Six Ranches are discussed in Chapter 3.

Sam Graves who rode Old Hub to brilliant victory at Haskell in 1898, faced the camera in this pose on Old Button just a half century later. See Chapter 2.

From a College Diploma to a 6666 Cowhorse. June Gibson on "Friendly." The picture was made on the 6666 Ranch in King County in 1947. Gibson is a graduate of T. C. U. See Chapter 3.

Chuck—at the 6666 wagon (King County) in 1947. The cowboys were too busy to pose. The names of the crew of cowhands were Jack Spencer, Riley Thacker, Toar Piper, Wayne Piper, Shorty Freeman, Shirley McClaren, Jake Luttrell, Ted Wells, John Dotson, Bill Mason, Latham Withers and J. J. (June) Gibson. This crew and Chuck Wagon made part of the story of Chapter 3.

The **FAMOUS BUCKSKIN MULES** of the Four Sixes—frequent prize winners. See Chapter 3 for the Four Six Ranch story in King County.

Jack Spencer snapped at the 6666 chuck wagon in 1947. Jack was wagon boss. See Chapter 3.

"Uncle Dick and Aunt Alice" Brannin. Both were held in high esteem by 6666 cowboys. See story of Uncle Dick's retirement in Chapter 3.

This old print was made on the 6666 Ranch in King County in 1926. Front row, left to right. Mutt Dobbs (on foot), Jim Key, George P. Humphreys (Present manager), Dan Loving, Lonzo Mayfield, P. I. Carter, Slim Whaley and John Stotts (present sheriff of Motley County). The last man to right on fence between second and third horse was Horace Bryant, wagon boss.

The **6666** Remuda on King County Ranch in 1947. The cowboys are catching up their horses for afternoon cow work. For names of the cowboys and many of the horses, see Chapter 3.

Photo by Ernest Lee

Top Hand of the Four Sixes — Miss C e l i a Ann Martin, granddaughter of Manager George Humphreys.

The Pitchfork Pets in 1929. Claude Flippin (left) with "Boots" and George Carter (right) with "Buckshot." Pitchfork Ranch story chapter 11.

These Pitchfork Pets were young wild horses in 1929. Claude Flippin (left) on Boots and George Carter (right) on Buckshot. The Pitchfork Ranch story—Chapter 11.

One of the Pitchfork Pets of 1929 made a fine cowhorse. "Boots", shown here with Claude Flippin, was later transferred to E. W. Hollar who was then a Pitchfork cowboy. Hollar rode the horse as long as he was employed by the ranch and was presented with Boots when he retired.

Calves at the Pitchfork Ranch (in Dickens County 85 miles east of Lubbock). Note the Pitchfork brand on one of the calves.

In the foreground Red Mud Lambert—one time boss at the Pitch-forks. See Chapter 11.

Coy Drennan—Pitchfork wagon boss on "W. P. A."

This horse known as Pitchfork Bay had quite a rodeo record. For the Pitchfork Ranch see Chapter 11.

Snapped on the old Bunk house steps at the Pitchfork Ranch in 1929. Left to right—Breck Jones, Rosie Deaton, George Carter, Riley Thacker.

Breck Jones on Badger (1929). Jones snapped many pictures on the Pitchfork Ranch at that time.

Riley Thacker on Ginger (Sept., 1929). This fiery Pitchfork horse had to be "broke" each spring for four successive years.

The horses were three of the first colts from the famous Pitchfork stallion, Blue Bull. Left to right: Red Bennett on Montie, Johnny Stephens, later Matador Ranch foreman, on Blue Bull and Wilburn (Rosie) Deaton on Sundown. The picture was taken about 1934 when Virgil Parr began to "breed up" the Pitchfork cow horses. See the Pitchfork Ranch in Chapter 11.

Old Print of the Pitchfork Ranch Remuda as it was twenty years ago—just before the "breeding up" process began. "You can't beat those old cowhorses," said King County sheriff E. W. Hollar, in sentimental mood, as he remembered his own days as a Pitchfork cowboy.

W. J. (Bill) Elliot—known to cowboys as Scotch Bill—came to America from Scotland in 1888 partly to look after the interests of Scotch stockholders in the Spur Ranch. He never lost his Scotch brogue but he became a good West Texas cowboy and a fine American. See Chapter 9.

Bill Stafford (late of Afton, Dickens County, Texas). He was a Spur Ranch cowboy in 1883 and for s o m e years afterwards.

The late John Gibson, retired f a r m e r and rancher of Paducah, Texas. He came to the area about 1882, and was a cowboy for the "JF" and "8" ranches.

Jim Gibson (of Grow, King County, Texas), prosperous farmer and rancher, came to the "8" Ranch in King County in the middle 1880's.

Lee Ribble (of V e r n o n, Texas). Cowboy and later manager for the Hesperian Land and Cattle Company (the old 3D outfit) of Cottle and King Counties as early as 1882.

J. T. Bond (late of Jayton, Texas). He was both a Matador and a Two Circle Bar cowboy. His days as a ranch hand began in 1882.

The late Jim Olds (of Crowell, Texas) with Stephens and Worsham (the old R2 Ranch) at present Chillicothe, Texas, as early as 1882.

Old "8" cowhands of King County, Texas, slightly more than 50 years ago. Note the high-topped boots. Their names from left to right: White Moore, Jim Gibson, Bert Lower, Claude Jeffers (great "bronc" rider). Dar Ratliff, King Sloan, M. Sloan (cook), "Lige" Hicks, and Press McGinty.—Photo courtesy of Jim Gibson, now a prosperous farmer and rancher of King County, Texas.

No glamorous 10-gallon rodeo hats! No fancy chaps! They wouldn't have a chance to make the movies dressed that way—but they were genuine cowboys of Texas and Oklahoma 65 years ago. Jim Olds, the central figure, was for many years R-2 cowhand of the Crowell-Chillicothe country. (See chapter 13.)

Outlaw Half Circle Steer

The old Outlaw of the Pease River Breaks. The late Jim Olds (left) and Henry Ferguson (right) posed with this cardboard dummy of the old steer that they had disposed of a half century ago. The exciting story is told in Chapter 13.

Coy Drennan — wagon boss at the Pitchfork Ranch in Dickens County, Texas. See Chapter 11.

THE WAGGONERS
AT WORK AND AT PLAY

*Out into the great Southwest came these
cattle kings of the long ago. They made an
indelible mark across the Texas landscape.
Milestones on their trail were Hopkins County
about 1845, Wise County about 1850, Clay and
Wichita Counties 1871 and Wilbarger County
about 1880. Great expansions came as they
moved herds up the trail in 1870 and as
they spread across Red River in the 1880's.
Daniel Waggoner, his son W. T., and his
descendents became one of the most important
cattle families in American History.*

*Their brand, the symbol of the ever growing
Waggoner empire, changed as they moved
along the trail—from D61 in Wise County
to D71 on the Wichita Rivers to the Three D's in
reverse (about 1881) on the Pease.*

*And now the hoof-beats of still other vast herds
tap out a message of new Waggoner progress
—of work but sometimes of play—inside
the great ranch gate south of Vernon.*

A typical, natural pose of R. L. More, Sr., with Waggoner's from 1900 until his death September 6, 1941, and first Manager of the W. T. Waggoner Estate. Mr. More was a famous bird egg collector, and in the picture is blowing bird eggs. His famous collection still remains at the business of his son, Robt. L. More Tire Company in Vernon.

Mr. R. B. Anderson, Manager of W. T. Waggoner Estate from 1941 until January 6, 1953, now Deputy Secretary of Defense, Washington D. C. Bob Anderson is a man with a future in National Affairs.

Photo Courtesy W. T. Waggoner Estate

Below: John Biggs, the present manager, is shown directing the affairs of the half-million acre Waggoner Ranch from his car radio.

Photo by Dick McCarty

Burning the D's into the cow's hide. G. L. Proctor, Wagon Boss, on horse; Henderson Wilkerson with Branding Iron; S. Propps on top of calf.

A Close Up. Harve Brothers with Branding Iron, assistant not identified.

Herding Cattle by helicopter on Waggoner Ranch.

But they still herd 'em on cow-ponies. Cowboys, let to right: Dick Harper, Earl Grubbs, Robert McElroy, Harve Brothers, Roy Wilkerson, Bunk Pippin and Eddie Browder.

It's Chuck Time at the Three D's. Cap Warren, Chuck Wagon Cook, (left) was the central figure in a recent Saturday Evening Post article.

The Big Truck Train ready to receive t h o u s a n d s of Three D cattle.

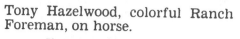

Tony Hazelwood, colorful Ranch
Foreman, on horse.

Cattle Loading Scene on the Giant
Waggoner Ranch.

At the Waggoner Loading Pens.

Ready to Go—The Waggoners were shipping them out.

The Boss and his wife—Mr. and Mrs. Johnny Biggs—in western regalia at the Big Show, The Santa Rosa Roundup.

Photo Courtesy W. T. Waggoner Estate

View of A. B. Wharton, Jr.'s Bath House. A. B. Wharton, Jr.—half owner of ranch.

Photo Courtesy W. T. Waggoner Estate

E. Paul Wagogner of Vernon, Texas, president of the Santa Rosa Roundup Association, on his famous parade quarter horse, Cowboy, with $3,000 silver-inlaid saddle. 'Widely-known Three D insignia of the Waggoner ranch is in the background.

Santa Rosa Roundup parade scene, Vernon, Texas.

pastures in a most lavish fashion by the Waggoners and Burnetts. The entertainment consisted of a unique wolf hunt[7] in April, 1905, four years after Burk Burnett and Joseph W. Bailey had asked the President for an extension of the lease. The Burnetts and Waggoners were still grazing 480,000 acres, or perhaps nearly half of their original holdings.

It is not unlikely that these large cattle barons still had hopes of extending their rights to graze this immense area in Indian Territory. It had been a very rich source of income to both of them and they would have been only human to wish for an extension. Not that any such motive would have been necessary to insure a full measure of Western hospitality from either Burnett or Waggoner, but it could have whetted their interest in the wolf hunt to a keen edge, never-the-less. Also they may have suffered some anxiety lest the President would be impressed with the possibilities of opening this Big Pasture for farming, and they may have wished secretly that the wolf hunt could be staged on some of their other lands. These interesting possibilities add a touch of color to the forthcoming wolf hunt. Just the presence of Theodore Roosevelt was enough to give color to the event without any such added features; and the hunter himself, who was destined to play the leading role for a half dozen thrilling packed days, was nothing short of a circus performer in his own right.

John R. Abernathy, commonly known as Jack Abernathy, was by 1905 one of the well known wolf hunters of the Texas-Oklahoma plains area. Through painful experience he had learned to catch wolves with his hands without the aid of weapons in a manner that had aroused the wonder of a great many people.

Jack had been a cowboy in West Texas during his boyhood, but in 1901 he had staked a 160-acre homestead some ten miles northwest of present-day Frederick, Oklahoma, in some of the pasture land that had been held by the Waggoners under a grazing lease. In the plain language of the West, Jack had become a nester, a classification that to say the least did not arouse the admiration of the cattlemen. Not only had he become a nester in the abstract—he had become one of the very people who had helped to break up this grass land area into farms.

To help maintain himself and his family, Jack became a peace officer even a little before he filed on the 160 acre claim. But re-

7. For the story of this wolf hunt see John R. (Jack) Abernathy, **In Camp With Theodore Roosevelt or the Life of John R. (Jack) Abernathy.** The Times Journal Publishing Co.. Oklahoma City, Oklahoma. 1933. pp. 100-127.

gardless of all else, Jacks' fame as a wolf hunter caused him to
be called occasionally to show his skill at wolf catching.

Late in 1904 Jack took part in a public program staged by
Cecil Lyon near Sherman, Texas. As one of the crowning events
of Lyon's entertainment program, Jack made a sensational ride
in the arena after a wolf, finally leaping from his horse, grabbing
the wolf by the under jaw, and bringing the animal to submission
in front of a wonder-struck audience.

Cecil Lyon, then Republican national committeeman from
Texas, was greatly impressed. Not long afterward Lyon was in
Washington and had an opportunity to tell Theodore Roosevelt
that he "saw a Texas cowboy catch wolves with his bare hands."
The national committeeman had to repeat his story to convince
the President that it was not a tall tale.

Roosevelt, who was soon to be Lyon's guest on a hunt to take
place in several southwestern states, wrote to Abernathy and
asked about the possibility of staging a wolf hunt. Jack wrote the
President offering to stage the hunt in the Big Pasture—the
480,000 acre grass lease mentioned above—and in accordance
with the plans of the presidential party suggested April 5 as the
date.

The arrangement was made and soon became known to the
public. Tom Burnett wired Abernathy to meet him in Wichita
Falls and talk the matter over.

The two men met in a saloon. Tom tried to persuade Jack
to move the hunt to the part of the Burnett pasture here east
of Electra, Texas. Finally young Burnett offered the famous wolf
hunter a thousand dollars to change the location. Jack who was
familiar with the lands in the Big Pasture refused. He pointed out
the danger in a fast ride through the mesquite covered lands near
Electra and insisted on the open country north of Red River as
the proper place. Hot words followed and the two men almost
came to blows, but no change in plans was made about the wolf
hunt.

It was 2 o'clock on the afternoon of April 8 when the presidential party arrived by special train from Vernon, Texas, at the
little new town of Frederick in what is now southwestern Oklahoma. Frederick had already acquired a population of 2000 inhabitants. It was located in a strip of the former Waggoner
leased pasture lands nearly twenty miles wide across the west
end of what is now Tillman County, Oklahoma. This was very
rich farm land that was already occupied by several thousand

families of homesteaders. It was part of the nester invasion that was already beginning to sweep across much of West Texas. It was a part of the greater westward farm migration that had seen the little band of Quakers huddle into their dugouts on the bleak plains of Texas twenty-five years earlier. Wherever this nester class invaded in forces, the ranches were soon to disappear.

Down in southwestern Oklahoma the Chamber of Commerce spirit had taken possession of the little town of Frederick. It was only six miles east from this new trading center to the 480,000 acre Big Pasture of the Burnetts and Waggoners. If that area could be opened up for settlement, it would accommodate 3000 more families of nesters with its consequent great enlargement of Frederick's trade territory. The President of the United States was on hand—the very person who in the last analysis could have most to say as to whether that pasture should be subdivided into farms.

Needless to say Frederick was draped in flags on this April afternoon. Several bands played and when Teddy stepped off the train a wild cheer went up—comparable to a cheer that now might result when the home team makes a touchdown on the football field. Roosevelt waved his hat and smiled as he boarded the carriage that was to take him two blocks away to a grandstand where he was scheduled to speak. A great cheering multitude had assembled ready to hear the President. Here were some thousands of the immense nester army that was invading the grass land empire, breaking up the big ranches with evident consequences upon the American supply of beefsteak.

The outstanding persons of the president's party were seated on the speakers' platform. They were Lieut. Gen. S. B. M. Young, U.S.A. retired; two former Rough Riders, Lieut. Fortescue, U.S.A., and Sloan Simpson; Dr. Alexander Lambert of New York; and the President's host, Colonel Cecil Lyon of Sherman, Texas. Also seated among these distinguished persons was the famous Indian Chief Quanah Parker and his three wives and baby. In addition to this interesting group on the speakers' stand were the two prominent cattlemen, Burk Burnett and W. T. Waggoner, who were to serve as local hosts to the President during the wolf hunt.

Quietly in the background of this gala occasion businessmen of the little town of Frederick undoubtedly had secret hopes of an increased trade territory—and just as intensely the two big cattlemen must have had hopes of quite a different kind.

The President began his speech while Jack Abernathy was

having the hunting party's camp equipment unloaded onto
wagons to make ready for the hunt. During the speech Jack rode
Sam Bass, his fine white horse, up to the grandstand and dismounted. When he came up the steps ready to take his seat,
Cecil Lyon interrupted the President long enough to introduce
Jack to the distinguished visitor. Roosevelt shook hands with the
wolf hunter amid the prolonged cheers of the audience.

When the President had concluded his speech, he and his
party drove eastward toward the scene of the hunt. It was six
miles to the Big Pasture and another dozen miles within this
cattle range down the valley of Deep Red Creek to the camp site
chosen for the hunting party. A little tent city was in readiness
by nightfall when the hunters arrived. It was to serve the presidential party on its unique wolf hunt during the next six days.

During the first morning out, Jack gave chase to two gray
wolves. He and Roosevelt led the chase over the prairie country
not far from the present town of Grandfield, Oklahoma. It required a swift run of about two miles before the dogs engaged
one of the wolves in combat. Abernathy dashed up and leaped
from his horse into the midst of the battle. With his gloved hand
he grabbed the wolf by the under jaw and without injury to
himself held the animal up so that the president could witness
the procedure.

"Bully," shouted Teddy as he glowed with the enthusiasm of
the chase. The hunting party was served its noon meal from
a chuck wagon that had been driven out on the range to a predetermined point. That afternoon the wolf hunter repeated his
performance of the morning with a second catch. The President
did not conceal his enthusiasm.

At night there was a bet between Al Bivens and Cecil Lyon
about the next day's wolf hunt. Bivens was to catch the next wolf
with his hands as Jack had caught them; but next day when the
test came, Bivens was badly bitten by the wolf and withdrew.
Dr. Lambert then stabbed the wolf to death with his knife, and
Cecil Lyon won the bet.

Later in the day Abernathy had another encounter with a
wolf. While the dogs held the wild animal at bay, Jack leaped
from his horse toward the fight and, to his surprise, the wolf
started to attack him. Nevertheless Jack grabbed the wolf's
mouth to catch his underjaw as usual and quickly made a cap-
tive out of the animal that had tried to fight.

By the end of the second day it was pretty evident to Cecil

Lyon that the cowboys and cattlemen of the hunting party had a strong feeling of resentment against Abernathy. He called Jack's attention to this feeling of animosity and suggested a test race for the third day—permitting each contestant to select two dogs in a race that pitted horse, rider and dogs against horse, rider and dogs to determine who could really make the catch. Jack consented to the arrangement although he had only his single horse, Sam Bass, against the daily change of mounts ridden by his opponents.

There was a little tenseness in the air on the third morning. After the members of the hunting party had ridden out in the open looking for a wolf, W. T. Waggoner taunted Abernathy as to why he hadn't started after a wolf that finally came in sight. Waggoner repeated the taunting remarks and a hot exchange of words between him and Jack came near the stage of a personal encounter.

Jack had dismounted and had taken hold of Waggoner's bridle reins when the President spoke rather sharply, requesting Abernathy to get on his horse. The matter was dropped, and a long chase for the wolf followed. Roosevelt waved his hat in the air to start the race.

Jack held his horse back for a while until the mounts of his opponents were "winded" and then he gradually gained until at last he took the lead. He caught the wolf after a nine or ten mile ride while nobody but the President was near him. Following from behind came Dr. Lambert, Bony Moore, and Tom Burnett.

The President extended hearty congratulations to Abernathy; and after the wolf was brought into the presence of the rest of the party, Waggoner also congratulated him. The race had been a great demonstration of endurance and skill on the part of Abernathy and his horse and dogs, and the entire party did not fail to recognize it. By the end of the six days he had caught fifteen wolves alive and had become the undisputed champion in that type of hunting.

Soon after the hunt, Roosevelt wrote his children complimenting Tom Burnett as a likeable person and exclaiming that Abernathy was wonderful. So great was the President's admiration for Jack's courage and skill that he appointed him State Marshal at a salary of $5000 per year—and that over the protests of many of the politicians. Later the President invited Jack to accompany him on the now famous big game hunt to Africa, but conditions at home made it impossible for him to accept.

Now let us turn back again to the wolf hunt. The whole six days seem to have been brimful of pleasure to Roosevelt; and perhaps without design, the trip incidentally constituted a tour of inspection of the Big Pasture for the President; and it should not be forgotten that the area was thrown open for settlement the following year. Bob McFall, late of Wichita Falls, Texas, who was a Burnett cowboy at the time of the big hunt remembered helping to drive the last of the Burnett cattle out of the Big Pasture in the fall of 1905.[8] The invasion by the plow was at hand.

Thus it was that within four years after the conference with President Roosevelt, all of the 6666 cattle had been moved out of Indian Territory, and the cattle empire of Burk Burnett had been moved to King County, 125 miles southwest of Wichita Falls and to Carson County, 30 miles northeast of Amarillo. Young Tom was not only a little more than thirty years old, but he was a veteran in the cattle business. He it was who went to King County to receive the newly purchased cattle acquired along with the old 8 Ranch—and to burn the now widely known 6666 brand into their hides.[9]

But Tom Burnett did not stay in King County. He had a niche of his own to carve out of the cattle world, and he came back to Wichita County—back to the ranch house on Red River— to begin to carve it. In 1905 when the last of the Burnett herds were moved out of Indian Territory, some of the cattle brought across the river were Tom's property. He had already begun his now famous Triangle Brand. He leased the old ranch in Wichita County[10] which was now far away from the center of the immense 6666 cattle interests. This little ranch had served as home base for the making of the Burk Burnett fortune—and it was now to become home base for the making of the Tom Burnett fortune.

In 1906, the part of the Burk Burnett Ranch that fronted on Red River was sold[11] and the headquarters were moved back to this place where my car is parked eight miles down the highway east of Electra. Tom and Ollie and their young daughter Anne, lived here until about 1912. The turn of events separated Tom and Ollie at or near that date, and they were divorced in 1919. Nevertheless the son of Burk Burnett was destined to go ahead in the cattle business.

Tom's Triangle branded cattle did not match in number the

8. Interview with Bob McFall.
9. John M. Hendrix as cited above, page 19.
10. John M. Hendrix as cited, page 19.
11. Interview with Bob McFall.

many thousands that by this time wore the four sixes, but there was slow steady growth toward the day when he could boast of great herds in his own name. In 1912 at the death of Captain Loyd, his grandfather, he inherited one fourth interest in the Wichita County ranch[12] and some money in addition. In 1918, the oil boom struck the town of Burkburnett; and even though the nearby portion of the Burnett Ranch had been sold, the sharp rise in oil lease values as well as some oil production on the land did something toward enlarging Tom's bank account.

But the fortune of the members of the Burnett family was soon to undergo the shock that must eventually come within the family of a multi-millionaire. Burk Burnett died in 1922,[13] and his will contained provisions that were to affect both people and localities for many years after his death. It prescribed some rules and regulations that were to govern a whole West Texas empire far into the future. So great was this will to influence the status of Tom Burnett that it is worth our while to turn aside here to consider the chief provisions of the important document.[14]

Perhaps it is not amiss to characterize this will as an instrument in writing full of the bigness of Burk Burnett. As a beginning of the long legal document $5,000 was set aside for a park in Fort Worth; $10,000 was left to a sister, Mrs. T. W. Roberts of Wichita Falls. Seven of Burk Burnett's nieces were to receive bequests of $5000 each and four of his faithful employees, Sid Williams, Bud Arnett, Joe Crystal and Charlie Hart, were to each receive a similar amount. William H. Hortenstein, a trusted friend and business adviser, was also to receive $5000, and Mrs. Margaret Pekor was to be granted $2500.

These bequests constituted indeed a large outlay of money, but they made hardly more than a dent on the Burnett fortune. The total of these lump sum bequests was only $122,500 out of an estate with an estimated value of $6,000,000 or about one dollar out of every fifty left by Burk Burnett.

In addition there were various monthly incomes provided. Miss Cora Belle Fuqua, who was an invalid, was left an income of $300.00 per month for life or at least during her illness. Sid Williams and Bud Arnett were each to have $100.00 per month if incapacitated for labor either by reason of advanced years or for

12. John M. Hendrix as cited, page 19.
13. The date of Burk Burnett's death is recorded in a court order: Wichita County Deed Records, Vol. 193, pp. 522-525.
14. Burk Burnett's will is recorded in Wichita County Deed Records, Vol. 201, p. 594 ff.

physical disabilities. Similarly John Stapp was provided with $30.00 per month and Burk Burnett did not even forget his old negro "Coley." Coley, whose real name was Oak Owens, was to receive $30.00 per month. Altogether these income bequests amounted to $560.00 per month or $6720 per year. This was equal to about one tenth of one percent of the value of the estate and of course did not constitute a heavy drain upon the income from the ranches.

About four times as large as all of these monthly bequests combined, was the $25,000 annual allowances which Burk Burnett left to his son Tom. The will denied Tom the right to share in the management of the estate.

Ollie was remembered handsomely in the will. She was to receive one-third of the net income after the several foregoing bequests were satisfied. Anne, Tom's only daughter and Burk Burnett's only grandchild, was to receive the other two-thirds of the net income.

Burk Burnett did not lodge fee simple title to any of his lands or buildings in anyone. He strongly requested Marion Sansom and W. E. Connell, his executors, not to sell the Burk Burnett or Reynolds buildings in Fort Worth, and also he requested that the large ranches in King and Carson counties be left intact. The executors were otherwise given the right to sell and re-invest. However, subject to any possible change by the executors, the various properties were to be held as an estate until the death of his grand-daughter Anne. At her death her child or children were to be vested with full title to this far-reaching fortune in lands and cattle.

There had been one miscalculation in the will of Burk Burnett. The rights of Mary Couts, his second wife, who had been adjudged insane were not considered important.[15] But events gave those rights far greater importance than Burnett had anticipated. Within about two years after the death of the cattle king, the courts declared Mary Couts sane again and it was only a matter of months until she had entered suit to break his will. Still short of a year she succeeded[16] and recovered some $3,000,000 worth of the great cattle fortune.

To give this surprising event another turn hardly expected outside of story books, the whole of this $3,000,000 worth of pro-

15. See Burk Burnett's will as cited above.
16. Wichita County Deed Records, Vol. 227, p. 319.

perty has since become endowment for a modern university.[17] This second wife of Burk Burnett set up the Mary Couts Burnett Foundation and gave this new institution possession of the sizeable sum just won in the country's law courts. She personally retained some of the income from this foundation, but this arrangement ceased at her death which occurred only two years later. Thereafter Texas Christian University of Fort Worth, Texas, became sole owner of this important source of income.

However, even though this deflection from the Burnett fortune was large and is due to scatter its rich benefits to many future generations, it made little more than a dent upon the story of the 6666 Ranch. The judgment[18] gave much personal property to Mary Couts—including, it is true, 4000 steers from the 6666 ranches that were, in all, stocked with some 20,000 head of cattle. Also an undivided half interest in the Carson County ranch was awarded in the judgment. But the 200,000 acre King County ranch, except for certain oil rights which soon expired, were not included; and only a part of the oil rights of the 25,000 acre ranch in Wichita County were awarded by the courts.

Thus the 6666 Ranch had only to lease the Mary Couts half of the Carson County lands to continue to do business at the same old stand without the loss of a single acre of grass. This is exactly what was done and the ranching business set up by Burk Burnett has continued in a manner not very different from that prescribed by the provisions of his will.

The Carson county ranch has produced many thousands of dollars from its oil and gas royalties both for the intended heirs and for Texas Christian University—but the 6666 Ranch goes on.

Burk Burnett's will has now been in force for more than thirty years. An actuary with a cold, calculating eye for statistics would probably expect Burnett's granddaughter to live at least another quarter of a century. Thus this cattle king, whose magnitude became apparent well before the end of the last century, has by the terms of his will prescribed rules of a little ranching empire that may last far toward the end of this century. There is an heir apparent to the throne of the far flung Burnett kingdom. Anne Valliant Burnett Hall, the only child of Burk Burnett's granddaughter Anne, is the young lady about whom news stories of fabulous wealth are most likely to be written when she becomes

17. Colby D. Hall, **History of Texas Christian University**. Texas Christian University Press, Fort Worth, 1947, p. 182.
18. Wichita County Deed Records, Vol. 204, pp. 33-48.

the actual owner of this wide spread fortune. The figure of **Burk** Burnett is still a living force in the cattle world of today and seems to project itself far into the **future.**

Tom Burnett and his daughter Anne seem to have worked out a somewhat different understanding between them from that left by the rather strict will of the elder Burnett. It seems that the will was not abrogated, but that Tom and Anne distributed their part of the income in a way that they believed was more nearly just to Tom than the rigid allowance left by his father.

Tom demonstrated his ability as a financier with this and other income that was available. He built up an estate in his own name that was to stand out as one of the large landed fortunes of West Texas.

It was 1923 when Tom began to grow into one of the great cattlemen of Texas.[19] Beginning in that year and extending over the remaining fifteen years of his life, he bought five ranches in Foard and Cottle counties and consolidated them into the one vast range mentioned earlier in our story. He bought the Pope Ranch in 1923, the McAdams Ranch of some 30,000 acres in 1924, and the Moon Ranch from W. Q. Richards in 1925. Four years later, he bought the 32,000 acre YL Ranch northeast of Paducah. It had been part of the OX Ranch that had been established around the forks of the Pease River by the Forsythes even before 1880. Nine years later, he bought 20,000 acres more that had once been known as the Dripping Springs pasture of the OX's. However, at the time of Tom's purchase it had become known as the 7L Ranch. The Triangle brand had at last spread until it covered quite a bit more than a hundred thousand acres, but the 7L Ranch was to be the last great addition. Tom Burnett, who had learned the cattle business in the 1880's and 1890's in the Indian country between the Wichita mountains and Red River, died in 1938, apparently in the midst of a long range campaign to build a fortune equal to that of his father. He fell short of that objective, but he grew in the cattle world as one of the pace setters of recent decades.

By his will[20] Tom left these extensive ranching properties to his daughter Anne, who was already the chief figure in the 6666 Ranches, thus making her one of the wealthiest owners of land and cattle known to the West. Both she and her young daughter

19. For an account of Tom Burnett's growth as a cattleman, see John M. Hendrix as cited, p. 20.
20. Wichita County Probate minutes, Vol. 30, p. 67.

bid fair to hold sway over the fabulous fortunes of the Burnetts for a long time in the future.

But while we are speaking of fabulous fortunes, let us go back for a few moments to the ranch which Burk Burnett established in Wichita County just south of Red River. This was not an enormous spread of acreage when compared with the great ranches. It was just three acres short of 17,000[21] but as it turned out, it was one of the most valuable ranches ever sold in North America. When Burnett—and Loyd who owned a one fourth interest— sold this bit of pasture land in 1906, it was a proper part of the well considered plans of a successful ranchman. The last of the Indian leases north of Red River, it will be remembered, expired in 1905. This 17,000 acres across the river had served its purpose as a base from which to operate the extensive Indian grazing lands. A good price—$289,000— was offered and it was the part of good judgment to sell. It so happened that the purchasers, J. A. Kemp, Frank Kell and others, were about to build a railroad across the ranch and open it up for farm lands. Even if Burnett had held his land and offered it for sale after the railway was completed, the sale price could not have been enormously greater.

As it happened, the purchasers laid out a townsite on a portion of the ranch lands in 1907 and subdivided the principal part of the ranch into 160 acre tracts suitable for farms. The town was named Burkburnett[22] in honor of the former owner.

Quietly, the new town and the new farm area about it settled down to the routine of daily living—farming and supplying the needs of farmers. Even before the breaking up of Burk Burnett's 17,000 acres of grass, a small farm community had sandwiched itself in between the two principal parts of the ranch. A rural store called Nesterville owned by J. G. Hardin had held forth a half mile south of the site of the new town for nearly thirty years before the coming of the railroad.[23] Hardin owned the greater part of a thousand acres of land that lay not far west and south of the new railway station. The Hawkins family owned the land that joined Hardin on the south and west. Together the Hardins and Hawkins owned land that cut a kind of sawtooth gap out of the

21. The deed to this ranch from S. B. Burnett and N. B. Lloyd to J. A. Kemp et al is recorded in the Wichita County Deed Records, Vol. 46, pp. 331-32.
22. The proper pronunciation of the cattle king's name, as well as that of the town that was named for him, was Burk BUR nett, not Burk Bur NETT. The name of the town was often correctly pronounced until about 1918 when the great crowds that rushed in to the oil boom town without any good reason began the present mispronunciation, Burk Bur NETT.
23. Interview with the late G. S. Hawkins of Frederick, Oklahoma.

south side of the ranch. To be more exact the owners of the ranch purchased land to the east, west and north of these families—for the Hardins and Hawkins and a few other families were in the country first. The elder Hawkins helped to build the large farm house east of Nesterville for Burk Burnett in 1883.[24] The old house still stands just south of the present-day concrete highway bridge that crosses Red River into Oklahoma.

For eleven quiet years the new farming area produced cotton and corn and wheat, and the little village of Burkburnett grew to be a town of a thousand or more inhabitants. Scattered areas of oil production had been developed to the southwest and also to the southeast of town.

Then one day in 1918, S. L. Fowler who owned a farm north of the little town proposed to drill a well on his land to test for possible oil production.[25] He organized a small company of 120 shares at $100 each so that the neighbors could help to bear the costs and share in the possible profits. The venture looked a little foolish, for he began drilling within less than a mile of a "dry hole." Some say that he went ahead with the project because his wife dreamed of great oil wealth spouting out of the ground.

Be that as it may, he continued down with the drill—down 1000 feet, 1200 feet, 1500 feet, 1700 feet. So far it was just a hole in the ground. But only a few feet deeper he struck pay dirt. The oil sand which he discovered looked too promising to believe. Casing was set and the well "brought in." It flowed at the rate of nearly two thousand barrels daily—and the Burkburnett Boom was underway!

Here on land that Burk Burnett had sold to Kemp, Kell and associates—and that they had re-sold to hundreds of settlers—the whole crust of the earth seemed to be only a cover to a vast pool of golden fluid. Financiers and promoters rushed in. They bought small leases and even town lots, expecting to drill for oil. Soon there were hundreds of producing oil wells. The new oil field soon reached eastward to within a half mile of the old Burk Burnett farm house and westward completely across the land of J. G. Hardin and a few hundred yards beyond reaching into the west part of the original 17,000 acre ranch. Millions of dollars' worth of new wealth came out of the ground; J. G. Hardin, already a man of some wealth, became a multimillionaire. Fowler and

24. Interview with the late Mrs. Hiram Willis of Wichita Falls, Texas.
25. The writer lived in Wichita Falls and Burkburnett before and during the oil boom that followed the discovery of oil on the Fowler farm.

his associates sold out for $1,800,000. Each of the neighbors who had purchased a $100.00 share received $15,000. Many of the other oil companies made rich profits, but many of the schemes of eloquent promoters left a trail of sadder but wiser stockholders.

The oil field proved to be about 4,000 feet wide and three miles long, with the town of Burkburnett in the heart of it. Soon a ring of dry holes bordered this rich producing area on every side and the Burkburnett Boom eased to a standstill.

But the great rush was not over. A new oil field was discovered in what had been Burk Burnett's big west pasture, less than a mile south of the place where the old ranch house had stood. It proved to be a much bigger pool than that discovered by Fowler. It was five miles long and two and a half miles wide, and every inch of it was within the original Burk Burnett Ranch.

I climbed an oil tank one day during this part of the boom and watched the flow from one of the wells. It was pouring into the tank through two four-inch flow-pipes—each gushing like a stream of muddy water spouting from the snout of a mad elephant! The well produced 5,184 barrels that day. At $3.50 per barrel this well alone would have paid back the $289,000 sale price of Burk Burnett's ranch within sixteen days!

So far as I know, this was the best well of the two pools, but there were literally hundreds of good wells. Had there been pipe line facilities, the total production of the two pools would have reached a possible peak of a quarter of a million barrels per day. It was not much more than a year until the combined output of the two fields had produced $100,000,000 worth of oil—and that large figure has been duplicated several times over during the long years that the fields have produced since then. Certainly more than ninety percent of this immense total came from the area that was originally Burk Burnett's 17,000 acre ranch. This cattle king of yesterday passed up one of the great fortunes of history when he signed the deed to Kemp, Kell and others back in 1906.

Instead he stayed within his character as ranchman until the end of his colorful career. Indeed he did amass a fortune of large proportions, but he made it out of a series of factors with which he had had much experience—cows and grass and cowboys; rain and drouth and blizzards; ever-changing markets, and an era of free grass that closed quickly behind him. He carved his fortune out of the rough unpolished elements of the old frontier and at the end of his day he loomed large among the men of the

old West. Had he kept the little 17,000 acre ranch and developed
it himself, he could have been one of the outstanding financial
figures of American history—not a great distance below Rocke-
feller and Henry Ford. But somehow I like his success story
better the way it is.

Burk Burnett lived in Fort Worth during most of the years
after he had become a big cattleman. He divided his time between
the great ranches of his estate and this modern Texas city.

Tom made his home in Wichita County for just about a half
century—all of his life after he first became a cowboy. This place,
the house eight miles east of Electra, was his home a great part
of that time. This was Tom's residence at the time of his death in
1938.

An interesting personality of the two large ranches, the Four
Six and Triangle, survived both Burk Burnett and his son Tom.
That person was Charlie Hart,[26] a one time employee of the
Indian Chief Quanah Parker. Charlie as a boy only nine years old
began work for the famous old chief and stayed in his employ for
a full half dozen years. At the age of fifteen the youngster began
his career as a cowhand for the Burnetts. Thereafter he never
knew any other master. Charlie received $5000, it wil be remem-
bered, set aside for him by Burk Burnett's will. Sixteen years later
he received $10,000 from Tom's will. But this was not the end of
Charlie's connection with the Burnett family. He served as Anne's
manager of the big Triangle ranch up in Cottle and Foard counties—served until sometime in 1947 when his retirement became
effective. He had worked for the Burnetts for fifty-six years when
his health began to fail. Anne offered Charlie a new car, but this
veteran of many West Texas winters refused it because his eye
sight was too poor to permit him to drive. She also offered Charlie
his full mount of something like a dozen fine cowponies, but the
old cowpoke accepted only three of them. There was no thought
of mercenary haggling over details in this transfer of parting
gifts; to the Burnetts, Charlie was like a member of the family!

But Charlie was much nearer the end of the trail than anybody realized. His death came not three years after his retirement.
More than almost any of his fellows he was strongly attracted to
the ways of the West. He could speak Comanche like a native,
but he also understood the language of the range so well that he

26. For the story of Charlie Hart, see Ernest Lee, Jr., "The Story of Charlie Hart,"
The Junior Historian, September, 1948.

knew by a cowboy's tone of voice whether his particular **brand of** profanity was the sharp challenge of an enemy or the **disguised** banter of a trusted friend.

Here at the ranch house of the late Tom Burnett I must **halt** and turn back into more of the great ranches.

15—The Greatest Ranches Today

No longer is there any such thing as a single ranch organization in the United States that owns a million acres of land—not even in Texas where the few largest ranches are concentrated. However, some surprises about large land ownerships probably await the reader of this chapter even if he has extensive knowledge of the big ranch country. Immense landed estates still border the map of Texas and one wonders how these great aggregations of land came into being in this particular part of the United States more than elsewhere. In part this condition originated with the King of Spain.

Perhaps it was beyond the powers of the average Spanish king to imagine any limits to his vast holdings in the New World. What if he did give away a spread of many leagues to one of his favored subjects—there was still more real estate extending all the way from Florida to the great South Sea. But long after the King of Spain had lost control in the New World, there were still forces that made for large land holdings. When Texas became a republic and finally a state, her lawmakers almost constantly faced the threat of an empty treasury. They also faced the problem of bringing immigrants into their sparsely settled country. Land was their best known answer to both problems. Land sales and land bargaining might fill their empty treasury. Once, when the question of locating the boundary of Texas came to the point of settlement, the United States government paid Texas well above $10,000,000 for 67,000,000 acres that includes much of present day New Mexico and other western states. For once, Texas could balance the budget with a surplus in the treasury.

The immigrant problem was also approached from the angle of bargaining. Railroads were induced to build into the heart of Texas by direct grants of as much as sixteen square miles of land for each mile of track. The premium offered by the state for 100 miles of railway was thus a little more than 1,000,000 acres of land. To be sure, such grants now seem excessive, but Texas in those days seemed like a mighty empty country, and the land had very little cash value.

Whether the land policy of the King of Spain or the Government of Texas was wise is now merely an academic question. The result of these great grants was to cover the State of Texas with many large bodies of privately owned land.

In some instances, such as the King Ranch and the Spur Ranch, all or part of a great tract of grassland was ready made for ranching. With the necessary cows, cowboys and cow-ponies added in each case, a ranch easily sprang into existence. As the corporate form of business organization grew in popularity, it became easy to get together sums of money sufficiently large to purchase these great blocks of land and to stock them with cattle. After the human imagination had been schooled to the successful operation of railroads and other big businesses. It is not at all surprising that the idea of bigness should have been applied to the cattle industry. The "beef bonanza" was widely publicized and much capital from as far away as Scotland came to the Southwest in general and to Texas in particular for investment in ranches. Human imagination has not yet suffered a major set-back in its capacity for bigness, but West Texas drouth and a crash in the beef market in the middle 1880's called a halt in the stream of money that had begun to flow into the big ranch country, and sober thinking held the more rampant imaginations in check.

Also there was the nester invasion, much talked of in previous chapters, which was a seemingly never-ending force toward the crumbling of the great ranching empires. But even the hardiest of nesters could not farm in the middle of the unspeakably rough Croton Breaks, nor could he make it rain on his plantings in the very dry country west of the Pecos River. Will Taylor, one of the large land owners of the Texas Panhandle, considers the area bordering the Canadian River as "ranchlands forever." His reason is very simple but full of logic. The land is too rough for farming. It so happens that most of the great ranches left in Texas are either in this type of country that is too rough for farming, or in the extreme western part of the state where scant rainfall is actually forbidding to the man with the plow. Without doubt, the nester invasion has gone much of the way toward its limit. To be sure, there are yet numerous acres whose sod can be turned into productive farm lands, but a new turn of events has called a halt.

Newer factors have come about that take the lead in determining whether a ranch shall remain large or crumble into lesser empires. The federal income tax throws its road block squarely in front of any big rancher who may wish to sell his property. The federal tax collector would take a heavy toll out of the sale, since much of his sale price in present-day low valued dollars would represent a paper profit. On the other hand federal inheritance taxes stand out as a nightmare to most of the big land owners, many of whom deed their lands to their children in parcels to avoid the day when the cost of passing a fortune on to the next generation proves so expensive in taxes. Thus it is that if a large ranch owner has three children, for instance, his great ranch, which cannot be sold without much loss in the form of income taxes, eventually (if gradually because of inheritance taxes) will become three ranches owned by his three offspring. Thus the lands of the big ranch country cannot in great measure respond to the law of supply and demand, but instead their course is kept in a straight jacket by federal taxes. Honest and expert study of this angle of federal taxation might serve to point out modifications in the present laws which would better promote the public welfare.

The big cattle empires, with their courses largely charted for them by federal taxation to begin with, are at present subject to another powerful force. Oil, the thing that has taken such a dominant position in the rural economy of most of Texas, has not missed the big ranch country. Many of the owners of large acreage are afraid to sell their holdings for fear some prospector will later strike oil on their land. Also, the bonuses or yearly rentals paid for oil leases are distributed throughout most of the ranch country. There are not many of the numerous great fortunes that have come to Texans in which oil has not in some way played a part.

Perhaps this brief survey has made it easier to understand why the big ranches have at least halted in their dissolution— but whatever has been the cause, we still have with us a number of ranches that are fabulously large. It is the further purpose of this chapter to pay each of them a visit, to take their measure, and to rate at least the few largest of them in their order of magnitude. This task is greater than it seems. I have driven several thousand miles through the big ranch country of Texas without fully covering the assignment. Only sketches of that journey can be related here, but the discoveries about present day cattle empires are all included in this chapter.

I soon learned to check the county tax records in order to discover the largest land ownerships, but my first contact was a much more pleasant experience than any mere turning of pages.

It was on my journey from Wichita Falls toward Amarillo, in the Texas Panhandle, that I interviewed W. J. Lewis, Jr. in the town of Clarendon. Lewis's father had purchased the Spur brand as told in an earlier chapter. Also, he had become owner of the historic Shoe Bar brand and in addition, the famous R O brand that had once belonged to the important Englishman, Alfred Rowe. The R O brand had from the early days of ranching in the Texas Panhandle belonged to Alfred and his two brothers. Their great ranch that lay eastward of present day Clarendon finally became the sole property of Alfred. Then in 1911 came the untimely death of this well known English gentleman. He went down on the great steamer Titanic in the almost unbelievable marine tragedy that shocked the world of that day. Later the elder Lewis acquired the brand and the remaining ranch lands of these noted Englishmen. Now, the Spur brand, the Shoe Bar and the R O (all three brands that have loomed large in Texas cattle history) are still worn by the cattle of the several Lewis ranches.

Young Lewis obligingly marked these various ranches on my map of Texas. The Spur and Shoe Bar brands are run on the Lewis ranches in Hall County south of Clarendon, and the R O brand is still burned into the hides of the cattle that now graze on part of the old ranch that was once owned by Alfred Rowe. The Lewis ranches altogether cover a total of 156,000 acres. Young Lewis also pointed out on the map the holdings of the well known Mill Iron Ranch—80,000 acres in Hall County and 20,000 acres in Collingsworth County in the east Panhandle. Historically this is, loosely speaking, the successor to the previously mentioned Millet Ranch which began in Baylor and Throckmorton counties even before the trail blazer Col. Goodnight had begun the first ranch on the high plains. It was in 1881, after six years of ranching by the Millets, that Hughes and Simpson purchased the Millet Ranch and for some years made it famous in the country near Seymour under their own Hashknife brand. Later Col. Hughes[1] of this firm of Hughes and Simpson moved his cattle investments up into the lower Panhandle and began these ranches that are now known as the Mill Iron Ranch. These and the other important ranches that must be mentioned all too briefly here are shown more accurately on the maps that accompany this chapter.

Some twenty miles up the road that extends from Clarendon toward the heart of the Panhandle is the colorful little town of Goodnight. Although it is just a village astride the highway, it stands as a lasting memorial to one of the most important and interesting characters of Texas ranching history—Col. Goodnight[2] himself.

The rugged westerner for whom this small town was named was the first cattleman to come to the Texas Panhandle. In 1876

1. In this brief note it is not possible to give a detailed account of the story of this ranch. In about 1872 Couts & Simpson began their large ranch (commonly called the Hashknife Ranch for their brand) in the east part of what is now Abilene, Texas. Colonel W. E. Hughes purchased the Couts interests in about 1880. In 1881 some 10,000 of their cattle were sent to the Pecos River since the Abilene County with its influx of immigrants had become a shrinking range. In the same year they purchased the Millet Ranch in Baylor County with some 20,000 head of cattle. The holdings were finally incorporated under the name Continental Land and Cattle Company, a $3,000,000 concern, with Hughes and Simpson each holding slightly more than 25 per cent of the stock. A number of large sales and acquisitions were made by the company. In the early eighties a ranch was established on the Powder River in Montana and in 1889 and during several years preceding, the company had acquired some 208,000 acres in Hall and adjoining counties about 100 miles southeast of Amarillo, Texas. The Pecos River ranch had been closed out before this time and it was this same year, 1889, that the ranch in Baylor County was disposed of. W. E. Hughes became the dominant figure in the company after their holdings became thus concentrated in the lower Texas Panhandle and now for many years the Mill Iron brand has held the place that was earlier occupied by the Hashknife.

Still more land was acquired. In 1896 the company purchased the lands of the old Rocking Chair Ranch in Collingsworth County in the east part of the Texas Panhandle. In more recent years the company has sold a great part of these lands to farmers and ranchers.

For a much more extended account of the Mill Iron Ranch, see N. H. Kincaid "John N. Simpson, Pioneer Cattleman and Financier" in The Cattleman, March, 1947. Also see Laura V. Hamner, Short Grass & Longhorns, 187-194.

2. See J. Evetts Haley, Charles Goodnight Cowman & Plainsman. The story of the J. A. Ranch begins on page 276 of this excellent volume.

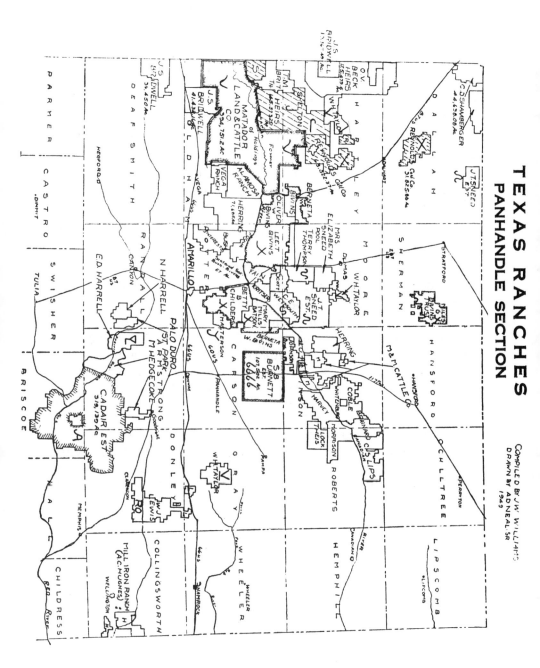

TEXAS RANCHES
PANHANDLE SECTION

COMPILED BY J.W. WILLIAMS
DRAWN BY A.D. NEAL, SR
1949

he drove his first herd of 1600 cattle to Palo Duro Canyon some thirty or forty miles southwest of the site of this village that now bears his name. He soon secured the financial backing of John Adair, Irish nobleman, and built up one of the greatest of all the big ranches. The initials of the Irishman, J. A. (connected), were adopted as the cattle brand of the new-born partnership.

In 1877 John Adair and his wife, Cornelia, accompanied Mr. and Mrs. Goodnight on the long 400 mile overland journey by wagon and horseback from Trinidad, Colorado, down into this new ranch in Palo Duro Canyon. The army had sent a detachment of soldiers at Adair's request, but even though military protection was at hand, the distinguished foreign couple had to camp out like all the rest of the party until a two-room log cabin could be built.

In order to experience something of the same feelings as did the Goodnights and Adairs, my wife and I once camped overnight alone in Palo Duro Canyon. We sat in the late evening and watched the sunlight and twilight play a medley of color against the red, yellow and brown rock formations, and looked 800 feet above us to watch the last faint tint of daylight die against the Caprock of the Plains. Probably it was not very different from the color drama seen in 1877—neither were the occasional bird calls that came in the night, but the parallel must have ended there. We drove out of the canyon some fifteen miles and ate breakfast in the nearby town of Canyon, for the High Plains in our day are blessed with the conveniences of civilization. When the Goodnights and Adairs camped in Palo Duro Canyon, it was seventy five miles to the nearest house that had been built for human habitation.

That the ranching venture undertaken by the strange partnership was highly successful has been told in an earlier chapter —also that the two partners after about ten years, in fact after John Adair's death, separated their holdings. Cornelia Adair kept the major part of the ranch including Palo Duro Canyon until her death and even now the ranch continues as the Cornelia Adair Estate. The thousands of white face cattle that now crop the grass in Palo Duro Canyon still are branded with the initials of the Irish nobleman, John Adair. However, the ranch no longer covers the 1,300,000 acres that its owners once controlled through land ownership and grass leases. But there are even yet 319,000 acres³ left of the great estate, and it is still one of the first dozen ranches in Texas.

Col. Goodnight in his later years with his fortune greatly reduced in size lived here just a quarter of a mile south of this town that bears his name. His old home, though in ill-repair, stands full two stories high—a place of character such as Col. Goodnight might have been expected to build.

The 10,000 acres⁴ that include this old ranch home are now a part of the Mattie Hedgcoke Estate, as are 40,000 acres⁵ of the former Cornelia Adair lands.

From this village of Goodnight I drove westward and northward, and in fact, zigzagged over most of the Texas Panhandle; but the details are too involved to relate. Let us transplant ourselves 100 miles west over at the Texas-New Mexico line in order to better review the remainder of the ranches of this part of the Texas Plains.

Once cattle ranged all over the Texas Panhandle, coming first to the broken lands that bordered its rivers and creeks and later to the flat lands above the Caprock. These flat lands were, as elsewhere, the first to go under the plow. In 1947, this farm belt of the Panhandle produced a record crop of wheat and the farmers cashed it in for prices that ranged as high as $3.00 per bushel. With bank balances bulging with wheat money and an eye to further profits, there is little reason to believe that much choice farm land in this part of Texas continues to grow grass for cattle. The cattle ranches have shrunk back into the broken lowlands to the area that in the main must remain as "ranchlands forever."

The Canadian River Valley is the heart and soul of this Panhandle grazing belt.

If we make an imaginary tour down this river eastward from New Mexico to Oklahoma we shall encompass most of the large Panhandle ranches. Here, at the point where that stream enters Texas, we come within the boundaries of one of the two great pastures that before the sale in 1951, belonged to the Matador Land and Cattle Company. This one tract of grassland contains

3. From the records of the Armstrong Abstract Co. (August, 1948) the C. Adair Estate in Armstrong, Briscoe and Donley counties held a total of 319,139.50 acres.
4. From the assessor-collector's office of Armstrong County (August 25, 1948) the acreage of this ranch was 11,467.9 acres. The old ranch house is shown in the photographic section of this volume.
5. The acreage was actually 41,140.53 acres (August, 1948) according to the Armstrong County Abstracting Co. See ranch maps in this volume.

a little more than 394,000 acres.⁶ East and west it stretches a
full forty miles. The Matadors own one or both banks of the
Canadian River for all this forty miles, and only the good Lord
and perhaps a few Matador cowboys know how many miles it
actually is by the meanders of the crooked Canadian.

It was Lee and Reynolds who pioneered as ranchers (with
their L E brand) on this part of the map. They had operated the
official store that served Fort Elliot, and they had run long freight
trains along the old wagon trail that connected that fort with
Dodge City, Kansas. Profits had piled up from this and even other
extensive operations until by 1878 the two big time merchants
began to spread their herds over this area that is now under the
fences of the Matador Ranch. Three years later Reynolds with
his brother bought Lee's share in the partnership and kept the
west end of this now great Matador pasture (and land that
stretched far to the north of the Canadian River beside) and the
L E brand. Shortly afterward Lee moved down the Canadian
and in partnership with Scott, a New York banker, started a new
ranch under the L S brand. Soon after 1900 Reynolds Brothers
sold their land to the Giant Prairie Cattle Company. It was not
the solidly blocked land that now belongs to the Matador. There
were some big blocks of land, but in part there were just strips of
land that bordered the creeks that empty into the Canadian and
ran along the Canadian itself—just the land that an open range
rancher would need to control the whole range. But soon the XIT
Ranch bought everything around these fractional holdings, and
it was only natural that deals should be made to consolidate the
properties.

In due time the deals came into being—but not with the
XIT as purchasers. In 1902 the Matador bought some 200,000
acres of this XIT land and awaited developments. In 1915 the
Prairie sold the L E 220,000 acres to J. M. Shelton. Soon the new
purchaser became dissatisfied with his property because the
Canadian River split it in the middle. He sold the half lying on the
south side of the Canadian to the Matadors and purchased an
equal amount of XIT land north of the river. Now his lands lay
in a single body—and the Matadors with other additions built up
their immense block to present proportions.

Most of the Matador holdings in this area came from land

6. The 1948 Tax records of Oldham and Hartley counties show that the Matador
Land and Cattle Company owned 356,362.28 acres and 38,388.92 acres respectively in
the two counties—a total of 394,751.20 acres. The records actually examined were
duplicates from the above counties filed in the comptroller's office at Austin, Texas.

that originally belonged to the giant 3,000,000 acre **XIT** Ranch—a smaller portion from the Lee-Reynolds lands.[7] The Shelton property originated from the same two ownerships. It is now owned and operated separately by the three Shelton heirs—Jimmie Shelton, Malcolm Shelton, and Martha Shelton Houghton.

But for some reason the great ranches along the Canadian River seem to sleep in an uneasy bed. Recently another change has come. The Scotchmen who for a long time have held more northwest Texas real estate than anybody else have sold their land and their cows. In August 1951, the giant Matador Land and Cattle Company went out of business.[8] After sixty-nine years of rain and drouth—of feast and famine—this last of the Scotch syndicates from 5000 miles over the sea retired from the western ranges.

The details of the transaction sound somewhat complicated, but the net effect is the breaking up of the immense ranch started by H. H. Campbell when (in 1879) he bought the rights of Joe Browning at Ballard Springs, a mile southwest of the present little West Texas town of Matador. In closely related movements an American syndicate took over the properties and, on paper at least, subdivided them into a number of smaller ranches. Some of these have been re-sold and new names put on the map.

North of the Canadian River and west of Channing, 47,000 acres have been purchased by Foy Proctor of Midland, Texas. Clarence Scharbauer, Proctor's friend and neighbor has purchased 67,000 acres across the river from Proctor's land and northwest from the little town of Vega. The Mansfield Cattle Company (Jack Mansfield, Montgomery Bros. and C. C. Wimberly, all of Vega, Texas, and Billy Curry of Garden City, Kansas) purchased 70,000 acres west of Scharbauer.

Down at the little town of Matador other changes in the big ranch have taken place. Some 20,000 acres off the south part of the old ranch that lies near Dickens, Texas, have been sold to local ranchmen. About 22,000 acres from the extreme northwest corner of the ranch have been sold. The largest single transfer went to the Rock Island Oil and Refining Company. Theirs is a

---

7. For the several deals and early facts about the lands of Oldham and Hartley counties that now belong to the Matador Land and Cattle Co. and to the Shelton heirs see Laura V. Hamner, **Short Grass & Longhorns**, 152, 153, 170, 171.

8. Many details of this sale were given in the **Matador Tribune**, June 12, 1952; and in the **Amarillo Daily News**, July 25, 1951. Most of the facts about the purchases from the big ranch at Matador were given by Fred Bourland of Matador, Texas.

121,000 acre spread that lies both southwest and northeast of
Matador. They have purchased both the dirt and the water that
were once under the feet of Ballard, the buffalo hunter; and of
Joe Browning, the early day rancher. They have acquired the
headquarters—beautiful ranch home and all—that is pictured in
the photographic section of this volume.

With the Matador Land and Cattle Company out of the run-
ing, a lot of the map of West Texas has been changed, but let us
return to the Panhandle Country northwest of Amarillo.

The Sheltons, the Matadors and their successors are just
the beginning of the story of change that has come over the West
Panhandle—the map has been almost completely made over since
1900. Far up in the very north end of Texas, reaching from the
northwest corner of the state as far eastward as the old Buffalo
Springs headquarters, C. D. Shamburger and family own 45,000
acres[9] of former XIT Ranch lands. Shamburger is a successful
lumber dealer who has expanded his investments into the ranching business. Some twenty miles southeast of Shamburger, the
Reynolds Cattle Company (not the same Reynolds who was Lee's
partner), owns 32,000 acres[10] of former XIT grassland. Another
twenty miles to the south the Reynolds Company has a second
pasture large enough to bring their total Panhandle interests to
95,000 acres[11]—and even this is a minor part of the Reynolds
holdings.

West of this second Reynolds pasture and reaching within
ten miles of the New Mexico line, W. H. Taylor of Archer City,
Texas, owns 32,000 acres[12] that once belonged to the XIT. Taylor,
with another ranch in Moore County and one in Gray County,
spreads over 80,000 Panhandle acres.[13] A careful check of brands
leaves little doubt that Taylor has kept alive the old V-Bar brand
of W. S. Ikard,[14] that covered the country twenty miles west of

9. From the 1947 Dallam County tax records. The correct total was 44,633.08
acres.

10. From the 1947 Dallam County tax records. The correct total was 31,825.88
acres.

11. From the 1948 Hartley County records, Reynolds Cattle Company has
63,522.37 acres in Hartley County. Its total in the Texas Panhandle is 95,348.25 acres.

12. W. H. Taylor holds 32,041.85 acres in Hartley County (1948 Hartley Co. tax
records).

13. A personal interview with W. H. Taylor revealed that he held some 16,000
acres in Moore County and 30,000 acres in Gray County and small tracts elsewhere.

14. Mr. Taylor purchased the V-Bar brand from a man by the name of Gillespia
who gave the V-Bar Springs as the origin of the brand. The V-Bar brand of the Ikards
(variously used on hip, side or shoulder) is shown in George B. Loving's old brand
book (pp. 48, 94) of 1881. The bar was under the V just as in the Taylor brand of
today.

the site of Wichita Falls more than seventy- five years ago. The Taylors purchased a herd of these V-Bar branded cattle not long after 1900, and the brand still remains as one of their prized possessions.

Most of the extreme west end of Hartley County, lying west of Taylor is the property of O. V. Beck and family. About thirty years ago Beck ran a bit shop in Nocona, Texas. The bit shop paid well, but the oil business promised even greater profits. He tried his hand at oil and won handsomely, and shortly there were new profits to be invested. He moved up into the Texas Panhandle and bought this large tract of land that stretched twenty-five miles along the New Mexico state line. Even after selling a large tract, he and his family still hold 125,000 acres[15]—acres all originally owned by Lee and Reynolds and the XIT Ranch.

The man who bought the tract of land from Beck was J. S. Bridwell of Wichita Falls, Texas. Like Beck's rapid rise, Bridwell's history is another of the fine success stories that still loom in this land of individual initiative and (somewhat) free enterprise. Something more than thirty years ago Bridwell was in the real estate business in Wichita Falls. One day he found a buyer for a town lot. The deal would net him $37.50 if he could swing it, but a very obstinate person held a claim that clouded the title to the lot. It required the full influence of Bridwell's banker and a cash payment of $2.50 (nearly all the real estate dealer had) to clear the title. The deal was made and J. S. Bridwell went ahead to greater conquests. Soon he put his efforts into the oil business. His was not the case of immense riches gained overnight. At one time he operated a "star machine", but with his rare financial ability and business judgment, he grew to be a big oil operator.

The ranch, recently purchased from Beck—19,367 acres—[16] was only a climax to his series of ranch investments. Twenty thousand acres in Clay County, about 20,000 acres that reached up Blanco Canyon from Crosbyton to the old Hank Smith place— and some larger tracts in the west Panhandle—these were the stepping stones that brought J. S. Bridwell up to the magnitude of a great rancher and to a well deserved directorship in the Texas and Southwestern Cattle Raisers Association.

---

15. These lands are shown on the 1948 Hartley County tax records as follows: David W. Beck 18,841.17 acres; O. V. Beck 51,587.72 acres; Mrs. Marilyn Beck Castleberry 18,169.70 acres; and Mrs. Margaret Beck Whitaker 18,011.20 acres.
  The 1947 Dallam County records show assessed to Mrs. Donald Bagot 18,869.61 acres. This makes a total of 125,479.50 acres.

16. The 1948 Hartley County tax records show 19,367.49 acres.

His two large tracts of ranch land in the Panhandle—36,000 acres and 41,000 acres—[17] came from the orignial XIT Ranch. They lie west and southwest of Amarillo.

Between these ranches and that Texas Panhandle city are the Alamosa Ranch, the Vega Ranch, and the Herring Ranch of 59,000 acres, 26,000, and 72,000 acres respectively.[18] Part of this great expanse of grasslands, like many of the other Panhandle landed estates, came from the falling fragments of the old XIT Ranch. As the map of today will show, not only are Lee and Reynolds, Lee and Scott and the great Prairie ranches gone completely from the Texas Plains; but the immense XIT Ranch, giant of them all, owns not a single acre of Panhandle real estate. Down at the south edge of Deaf Smith County and in Parmer County below the Panhandle proper, this one time 3,000,000 acre titan of the cattle world owns a paltry 20,000 acres[19] of the great spread that the State of Texas once paid for its impressive capital.

The XIT has been out of the ranching business for more than forty years; now it is nearly out of the land business. New faces have come into the West Panhandle—farmers on the smooth tillable plains, ranchers along the broken valleys of the Canadian and its tributaries.

Eastward down the Canadian the changes have been equally pronounced. Nearly fifty miles downstream from New Mexico was once the little wild west town of Tascosa. It was the center of life in quite an area of the new Plains country seventy-five years ago. On the north side of the river as early as 1877 adjoining this town were Major Littlefield's cowboys and his 10,000 to 12,000 head of cattle bearing his famous L I T brand. East of Littlefield and on both sides of the river were the somewhat extensive herds of Bates and Beals with their L X brand. These two open range ranches filled a great part of present-day Potter County and at places spilled over the edges a little. Littlefield sold to the Prairie Cattle Company in 1881[20] and Bates and Beals sold to the American Pastoral Company in 1884.[21]

17. The correct figures from the J. S. Bridwell office records Wichita Falls, Texas, are 36,450 acres and 41,431 acres respectively.

18. From the 1948 tax records of Oldham and Potter counties the exact figures are 59,491 acres, 26,048.9 acres and 71,685.8 acres respectively.

19. J. Evetts Haley, **The XIT Ranch of Texas**, (University of Oklahoma Press, Norman, Oklahoma, 1953), 225.

20. J. Evetts Haley, **George W. Littlefield Texan**, (University of Oklahoma Press, Norman, Oklahoma, 1943), 129.

21. John Arnot "Famous LX Ranch had its beginning on the Arkansas River" in the Amarillo News Globe, (August 21, 1938), Sec. E., p. 22. John Arnot, an old LX cowboy, wrote the story of the LX Ranch for this Golden Anniversary edition of the News Globe.

It was the same year as the Littlefield sale that Glidden and Sanborn, barb wire kings, bought and soon fenced most of the part of Potter County that lies west of the site of Amarillo and south of the river into another interesting Panhandle ranch.[22] The truth is, the owners intended to call it "Panhandle Ranch", but just try to poke any expression down an old cowboy's neck if something else seems a little more natural to him. For example, Rev. L. H. Carhart who began his religious colony among early settlements of the Panhandle thought it quite nice to name his place Clarendon in honor of his wife—but the cowboys dubbed it "Saints Roost." The early ranchers of Garza County, southeast of present Lubbock, called a certain muddy little stream Dalton's Creek. When the Spur Ranch took over that part of the range, the Spur cowboys with full realism called the small watercourse "Slicknasty Creek."

The cowboys on the new ranch in Potter County were no different from the others. When their brand became known to the ranch hands—a pan with a rather prominent handle—they called it plain frying pan and the ranch went down in history as the Frying Pan Ranch. This big cow pasture enclosed by a wire fence demonstrated rather forcefully the practicability of barbed wire. It was not long until barbed wire enclosed nearly every body of land that was worth enclosing. It was through the influence of the Frying Pan Ranch ownership that Amarillo (then called Oneida) won, when there was a contest to see which of two new villages should become the county seat of Potter County.[23]

Perhaps the most outstanding event of its kind during the last half century of Panhandle cattle history was the rise of Lee Bivins as one of the big figures of the cattle world. The most important milestone on his sensational rise came in 1910 when he purchased 100,000 acres lying just northeast of Amarillo from the American Pastoral Company.[24] Bivins did not stop until he had acquired the range on which Major Littlefield once grazed his cattle. He even spread this purchase until it enclosed the ruins of the little old rowdy cow town of Tascosa and extended northward far enough into Hartley County to surround Channing, its modern county seat. In time, the Bivins lands reached still farther afield to include the large Coldwater Ranch in east central Sherman

22. John Arnot "Horses and the Frying Pan," Amarillo News Globe (August 21, 1938), Sec. E, p. 27.

23. Texas A Guide to the Lone Star State, 161.

24. John Arnot, "The Famous LX Ranch had its beginning on the Arkansas," Amarillo News Globe, (Aug. 21, 1938), Sec. E, p. 23.

County seventy-five miles north of Amarillo and another giant
sized piece of real-estate near Fort Sumner, New Mexico. In 1929
when Lee Bivins died, he is said to have owned outright a half
million acres of land and to have had a leased acreage that generally exceeded in area the land to which he held title.[25]

One of his valued possessions, rare as an heirloom, was his
ownership of the LX brand that Beals and Bates had brought to
the Texas Panhandle in 1877. Passed down through the American
Pastoral Company to Bivins and from Bivins to his heirs,[26] this
brand still stays on the hides of many North Texas cattle. More
than 400,000 acres of land[27]—no longer ranched as a single unit—
are held by the heirs of Lee Bivins.

Another forceful figure in Panhandle ranching history was
R. B. Masterson who, together with his sons, purchased the
American Pastoral Company holdings of about 100,000 acres[28]
north of the Canadian River in 1911. Previously he had gone to
Greer County, as early as 1882, and had moved up into the east
part of the Texas Panhandle. Then, in 1898, he acquired about
100,000 acres in King County, and the partnership "R. B. Masterson and Sons" had its beginning. When this partnership bought
the land from the American Pastoral Company, they owned some
200,000 acres and after such an expansion entered the ranks of
the great ranchers of the Southwest.[29] The property was divided
in 1917 and since has undergone still further subdivision among
the various heirs.

Overlapping now with a part of these acquisitions of R. B.
Masterson and sons are the holdings of some 50,000 acres in
northeastern Potter and southeastern Moore County by C. E.
Weymouth and son.[30] Adjoining Weymouth to the north, J. T.
Sneed and family own some 70,000 acres in Moore County.

If we add to this land a smaller ranch in northeastern Dallam

25. Dick Breen, "Lee Bivens, Giant," **Amarillo News Globe,** (Aug. 21, 1938) Sec.
E, p. 7.

26. The exact and original LX brand is the property of Mrs. May E. Bivins, the
surviving wife of Lee Bivins. Variations of this original brand with the "L" extending
from different corners of the "X" were run by the late Miles G. Bivins and by other
members of the family.

27. Listed in some ten different ownerships (some of which are most likely overlapping) in Potter, Oldham, Hartley, Moore & Sherman counties the Bivins heirs held
a total of 398,182.94 acres according to 1938 tax records. Small remnants of the property in Carson and Hutchinson counties are sufficient to run the total above 400,000
acres.

28. Now this property is divided in some five different ownerships. Part of the
land now belongs to C. E. Weymouth a son-in- law of R. B. Masterson and a recent
president of the Texas and Southwestern Cattle Raisers Association.

29. Gus L. Ford, **Texas Cattle Brands,** 214.

County, the Sneeds still hold something near 100,000 acres.[30] U. S. Highway 287 northward from Amarillo skirts the west side of the Bivins lands for more than fifteen miles out of that Panhandle city. Across the Canadian River it crosses the former Masterson lands and touches the Sneed property.

Down the Canadian River in Hutchinson County are the oil-rich lands of Johnson Brothers, Jessie Herring Johnson, G. A. and J. A. Whittenburg, J. J. Perkins and R. B. Harvey—none of them in the extremely large ranch class as total acreage goes, but most of their holdings very valuable when oil is considered. The largest acreage held in the county belongs to W. T. Coble and his daughter Catherine Coble Whittenburg. They have a total of 63,000 acres[31] in one of the most interesting of modern Panhandle ranches.

Here, bordering the Canadian on the north and down stream from present day Borger, came Thomas S. Bugbee, the first Texas rancher to locate in the Canadian River valley. He was only a fraction of a year behind Col. Goodnight who, together with John Adair, established the immense ranch in Palo Duro Canyon. In 1878 less than two years after Bugbee, came R. S. McNulty with his Turkey Track brand and located in this same area to the northeast of modern Borger. McNulty sold to C. W. Word in 1881, and the next year the Hansford Land and Cattle Company bought out both these early ranches and adopted McNulty's Turkey Track brand. The Hansford Company kept these holdings and a large acreage nearby for more than twenty[32] years but began to sell out piecemeal soon after 1900.

W. T. Coble, who began in 1899, did not have enough capital at first to purchase such a ranch, but he grew with the years and now he and his heirs own the lands on which both Bugbee and McNulty began in the Panhandle. The stone ranch house built by Bugbee, modernized but with its pioneer atmosphere preserved, is a part of this ranch—also the Turkey Track brand of McNulty has been revived, and two old ranches live on under a new setting and under a new ownership.

South of Borger lies the 107,000 acre[33] ranch mentioned in

30. Interview with C. E. Weymouth.
31. In the three names W. T. Coble, Catherine Whittenburg and Catherine C. Whittenburg, guardian, the 1948 Hutchinson County tax records show a total of 63,217.9 acres.
32. For the story of the Turkey Track Ranch see Laura V. Hamner, Short Grass & Longhorns, 131-140.
33. Wichita County Deed Records, Vol. 424, pp. 33-38. The Dixon Creek Ranch was 107,502 acres.

a previous chapter which Burk Burnett established on Dixon Creek. Sightseers have easy access to it, for a paved highway divides it almost in the middle.

From the foregoing pages it has become evident that although many ranches remain in the Texas Panhandle, the process of subdividing among heirs that has gone on in recent years has reduced the average size of a ranch materially. Because of this trend, we find the parts of the once great Bivins Ranch too small to enter among the dozen or more great ranches of today. The Shelton and Masterson holdings are likewise broken up among the members of those families until no single unit spreads far enough to enter into the final summary of present day giants to appear toward the end of this chapter. From this extreme northern end of Texas only the Matador (before 1951), the Burnett, the Adair and the Reynolds ranches are large enough to enter that final reckoning—and all of these except the Adair have greater holdings elsewhere. We have already had occasion to visit the principal unit of the Burnett Ranch in King County and the chief pasture of the Matador Ranch in Motley County. Both of these lie in the Wichita Falls-Lubbock country nearby but just out of the Texas Panhandle. In the same area with these two immense properties, we have already noted the Swenson and Waggoner Ranches which also rank among the great cattle companies of their day. The principal pasture of the Reynolds Cattle Company lies in extreme western Texas and still ahead of this my final journey through the big ranch country.

At the end of my stay in the Texas Panhandle I drove southward out of Amarillo to the town of Canyon, across the old T-Anchor Ranch of other days and southwestward to Hereford, at the gateway of the one time giant XIT Ranch. No longer is any of this flat plain country a part of the big ranch district. As mentioned earlier, the XIT Ranch itself, once three million acres strong, now holds only about 20,000 acres. That little remnant of the greatest ranch in American history lies southwest of this town of Hereford toward New Mexico partly in Deaf Smith and principally in Parmer County. The XIT cowboys ride no more except in memory; the remaining lands of the great ranch are now nothing more than marketable real estate.

My journey southward continued over prairies as flat as a pancake—prairies that are now mostly cultivated farm land. I drove through Dimmit and on south to the village of Springlake, some ten miles east of the old XIT Spring Lake headquarters. This village itself is in an extensive farming district but is not without its reminder of the big ranch country. The Mashed O Ranch of almost 75,000 acres[34] belonging to the Halsell Cattle Company lies to the west and south of the town of Springlake. Their headquarters rest in an extensive grove of trees near the same spot that was once the XIT Spring Lake headquarters. Ewing Halsell of San Antonio, Texas, one of the owners of this Spring Lake ranch, is also the owner of a ranch near Vinito, Oklahoma and more than 90,000 acres in Dimmit and Maverick Counties, in Texas.

From this village with its expanse of farms and its suggestion of big time ranching, I continued my drive southward. At Littlefield some twenty miles on my way, I met George White who owns the remaining acres of the former large ranch established by Major Littlefield, and his two nephews Tom and Phelps White. George, the son of Phelps White, lives in a beautiful home at the southern border of the town of Littlefield. It was through his courtesy that his ranch house by the side of Yellowhouse Lake is included in this volume.

Next in this trek across the flat plains of western Texas was the town of Levelland which it will be remembered was first laid out by C. W. Post under the name of Hockley City. Levelland is larger than Post's own Post City partly because of its greater trade territory in the South Plains farm belts and partly because of the discovery of oil a few miles away. The old Slaughter Ranch west of town has been subdivided and most of it sold, and the Spade Ranch which once formed a long strip of land almost across the east part of both Lamb and Hockley counties has dwindled to a remnant of less than 25,000 acres.[35]

Flung far to the right and left of the next 130 miles of my southward journey are three or four bodies of land, only one of which approaches the size of the great ranches of Texas. West of Tahoka in Lynn County is the 60,000 acre[36] property of the Crawford Estate and C. O. Edwards. Somewhat smaller is the C. S. Dean Ranch[37] (once owned by C. C. Slaughter) in the broken country east of Lamesa and the Higginbotham Ranch[38] in the

34. The 1948 Lamb County tax records show the Halsell Cattle Company with 72,935 acres within Lamb County.

35. The 1948 tax records of Hockley County show 21,754.02 acres.

36. The 1947 tax rolls of Lynn County show 58,488.5 acres.

37. This ranch is chiefly in Dawson County but extends eastward into Borden County. According to the 1947 tax rolls of the two counties the total acreage is 38,078.49.

38. Gaines County tax records for 1947. The Higginbotham Cattle Company had 52,467 acres in the county.

northwest corner of Gaines County. Quite a bit larger than any of these is the 167,000 acre Faskin Estate in southeast Andrews and adjoining counties. More than 100 square miles joining this tract on the east belong to J. E. Mabee of Tulsa, Oklahoma. These latter two tracts of land once constituted the C Ranch which Nelson Morris[39] came to West Texas and opened in 1883. Foy Practor of Midland at the time of this journey had a grass lease on both the Faskin and Mabee lands and in partnership with his brother, Leonard, he owned an additional 50,000 acres in the Holbrook country of Arizona.[40]

My route into this section took me from Levelland through Brownfield and on down to Seminole where I spent the night. Arising early the next morning, I pointed the nose of my car toward Odessa, sixty-two miles to the south. In this early morning trek my nostrils sensed one of the dominant forces that operate in the Texas ranching country. Everywhere the smell of oil added its unpleasant tang to the morning air. Oil is the great turn of good fortune that most ranchers hope for. It is the thing that makes most ranchers afraid to sell their lands for fear of giving away a fortune.

Back of this cautious attitude is the cold fact that Texas with 169 million acres of land has produced some sixteen billion barrels of oil—or roughly ninety-five barrels of oil for every acre of land. At current prices this startling total would amount to something near $260 per acre. The experts appraise the known reserves of Texas oil not far below another fifteen billion barrels[41] which means that we may estimate the total oil already produced or in sight at nearly $500 for every acre of ground in Texas. One eighth of this $500 or a little more than $60 per acre is the land owner's share known as royalty. How much future oil discoveries will increase this figure, of course, is mere speculation, but it is enough, together with the known facts, to make a rancher pretty conservative about selling his land. A 40,000 acre ranch, for example, with average luck will some day produce well above two million dollars worth of oil royalty to the land owner. At the same rate the 200,000 acre Burnett Estate in King County, which since this journey was made has begun to produce oil, has a potential of

39.   Gus L. Ford, **Texas Cattle Brands**, 54.
40.   Interview with Foy Proctor at Midland, Texas, in 1948.
41.   **Texas Almanac, 1954-1955.** In this volume of the Texas **Almanac** (p. 237) the total proven oil reserves in Texas amounted to 14,916,000,000 barrels. Also the same volume (p. 230) recites 15,759,500,000 as the total production of Texas up to the end of 1953. The price ($2.70 per barrel) upon which oil values are calculated here, comes from the same volume (p. 233).

more than $12,000,000 in future royalties. The former Matador Ranch, spread out from the head of the Wichita to the Canadian River, has a potential of not far from $50,000,000.

Now probably a good geologist would laugh till his sides hurt at this indiscriminate valuing of potential oil wealth. With his seismograph, torsion, balance and other scientific aids he can earmark certain spots on the map where the chances for oil production are a dozen fold better than just average. But some of these spots that have shared his richest blessings have failed to produce enough oil to soil your hat band, while others upon which he has frowned darkly have produced a fortune between suns. However, after all discrediting experiences are taken into consideration, one must admit that the findings of a geologist greatly enhance or greatly depreciate any particular spot of ground in so far as its potential oil worth is concerned.

If geologists had carefully located all the oil structures in the big ranch country and had published comprehensive maps showing favorable and unfavorable zones for exploration, such maps should have an important bearing on land values—but alas, not all of the country has been so mapped and only a small fraction of the information already known to a few big oil companies is in the hands of the general public. Hence, it is that the average rancher is none too anxious to sell the dirt out from under his feet for fear that dirt is just above an oil pool. Even if some oil company knows either favorable or unfavorable geology on his particular piece of earth, he can usually wait until the drill solves the riddle.

In practice, the discovery of oil has not occurred in any direct proportion to the size of ranches. Far down to the south of the Midland-Odessa country, Ira G. Yates, with little more than 20,000 sun-baked acres nestled down against the west side of the Pecos River, has from a discovery made on his land in 1927 reaped one of the greatest oil fortunes[42] ever to come to a cattleman. By contrast the Matador Land and Cattle Company with forty times as much land at the time of the sale had not received its first barrel of royalty oil. But the canny Scotchmen kept half of their royalty with an expectant eye toward the future. The great King

42. Carl Coke Rister, **Oil! Titan of the Southwest**, 292 ff.

Ranch of South Texas has been touched by the hand of oil but not nearly so bountifully as has this little (by comparison) 20,000 acre Pecos River ranch.[43]

Up the river from Yates and some thirty miles south of Odessa, the McElroy Ranch Company reaped some of the profits from West Texas royalty oil. Still farther north the Scharbauers of Midland hotel fame were touched by the end of oil's rainbow on their ranch northwest and later their ranch east of Odessa. Over in Winkler County the heirs of the colorful Bill Scarborough have their share in the great Permian Basin oil reserves. All three of these last are owners of ranches that range upward of fifty or even upward of hundred thousand acres—large by usual standards but still small compared with the few Texas giants.

An interview recently held in Odessa with tall, somewhat serious looking Marvin Henderson and rather short but fun-loving Cal Smith told of the good fortune that had come to a number of other middle class ranchmen. Both of these men were ranchers who had struck oil. Henderson, whose oil rich land lay to the southwest of Odessa, told how the drill bit had brought new wealth to his neighbor, R. T. Waddell, and had caused quite a spread of oil derricks on the fifty section ranch of the Connells a little farther to the north. Smith's land is in the North Cowden Field north of Odessa. S. B. Wight, his neighbor, and the J. L. Johnson Estate are extensive royalty holders in the same field. West of them in and near the northwest corner of the same county Pete Wheeler and Paul Slater share another oil producing area with Mrs. Thomas in this never ending Permian Basin country.

According to Henderson and Smith, the Walden and O'brien ranches over in Winkler, the next county west, were reputed at one time to have possessed more oil wells than any other land ownerships—but records fall pretty fast in the limitless Texas oil fields and nobody knows who may hold that distinction now.

When the conversation paused for a moment, I asked Cal Smith how the Cowden oil fields had happened to be so named.

"Because the discovery was made on Dick Cowden's land here fifteen miles north of Odessa," was Smith's explanation.

"Was this the same Dick Cowden who had a slight growth on one of his eyes," I wondered.

43. For oil production statistics by Texas counties see the **Texas Almanac 1954-1955**, 232-233. For statistics by oil fields see Carl Coke Rister, **Oil! Titan of the Southwest**, 408-415. Study of these statistics makes some deductions possible about individual ranches.

It was, Smith told me and thus gave me an interesting personal angle about fortunes in Texas oil. I had known Dick Cowden as a good natured school boy in Midland more than thirty years ago. He seemed too carefree and happy then to be forced to worry about the investment of any considerable sum of money, but since then he had acquired some three thousand acres of land north of Odessa—and now every acre of it is producing oil!

And so it is—oil has placed its touch of prosperity far and wide for two or three hundred miles up and down this Pecos River and South Plains country. It has enriched land owners by the dozens and has, for the time at least, disappointed others by the hundreds. From the standpoint of the unschooled public it has touched without rhyme or reason.

Up in the middle of the Texas Panhandle to the northeast of Amarillo it has run its golden path across several of the medium-large ranches, but it has touched only one of the giants—the Burnett Estate mentioned in an earlier chapter. Oil has so far eluded the several giant or near giant-sized ranches west of the Pecos River. Most of the extremely large ranches have been slow about taking their places as large oil producers.

There is one noted exception. The half-million acre ranch now owned by the heirs of W. T. Waggoner was the site of the discovery of the Electra oil field sometime before May 1910.[44] Earlier Waggoner had drilled several wells for water, only to find the wells polluted with oil. According to a local story he rammed a post down into one of these wells and abandoned it with disgust. He failed to produce good stock water, but this oil strike made sometime before the summer of 1910 was something new to think about. Only about a year after this discovery on the Waggoner Ranch, the Clayco Oil Company brought in a gusher near by and brought excitement at Electra to fever heat. The Electra Field was an important discovery, but the greater part of it lay to the east of the Waggoner Ranch. It was not until 1920 that oil was

44. The exact date of the discovery of the Electra oil field is not known to the writer, but the evidence that one or more commercially producing oil wells were in existence near the present Waggoner refinery before May 1910 is hardly subject to question. The late Joseph T. (Cotton) Young, of 1508 Monroe Street, Wichita Falls, Texas, was night driller on the Texas Company's No. 4, W. T. Waggoner, during the spring of 1910. He remembers this fact about the date because he saw Halley's Comet in the sky from his rig at night. While he was drilling this No. 4 well, he remembers that the No. 1 well was pumping an amount not in excess of 30 or 35 barrels per day. The No. 1 well had been drilled by Blackie Call and Earl Slater with a cable tool rig. Young's well was not a commercial producer but No. 5 which seems to have been completed in January 1911 proved to be a paying well. The Clayco No. 1 well on the nearby Putnam farm (of April 1911) was the first big well in the Electra field.

discovered in the heart of the big ranch itself. Since then most of the oil produced in Wilbarger County has come from the Waggoner Ranch—and Wilbarger County has produced a total just short of 110,000,000 barrels of oil![45] The King Ranch and The Burnett Properties are making great gains but are not up to the Waggoners as yet.[46] It is on the smaller ranches near Snyder and in the plains and Pecos River country where the fabulous fortunes have been made.

The C. C. Slaughter Ranch, which was once very large, unlike the Waggoner lands, had undergone much subdivision before 1937, when the discovery of oil in Hockley County came to some of the Slaughter heirs who still held a part of the old ranch.

I had passed only five miles east of this Hockley County field, a short distance south of Levelland, on my way to Seminole and Odessa. But despite its magnitude, I had not noticed the smell of oil—and not for many miles ahead was it forcibly brought to my attention. However, from Seminole southward the morning breeze was pretty heavily laden with the scent of that new evidence of wealth. This area is still part of the Texas ranch country, although none of these ranches as already intimated are among the giant spreads of acreage such as may be found in other sections. To be sure, I did travel eight of the sixty-two miles between Seminole and Odessa in crossing one end of the old "C" Ranch. But the most noteworthy blocks of land encountered during this short, early morning drive were not privately owned.

Eighteen miles on my morning journey I came between two large tracts belonging to the University of Texas. One of these, four miles to my right, was seventy-four square miles in area. The other tract, four miles to my left, had an area of 228 square miles. I traveled ten miles to the town of Andrews mostly between these two bodies of land. It was only five miles south of Andrews that I entered another great block of University of Texas land—this time 167 square miles in area. It was only eight miles across this long, narrow body of real estate; but that, together with what I had already encountered, set me off into a rather interesting chain of thought.

Early in its history Texas had set aside a million acres of fine, well-watered lands with the intent of financing a great University. Many leagues of the land were chosen in Grayson

45. Texas Almanac, 1954-55, 233.

46. The production in counties in which these ranches are located has boomed greatly during the last few years. See the Texas Almanac, 1954-1955, 232-233.

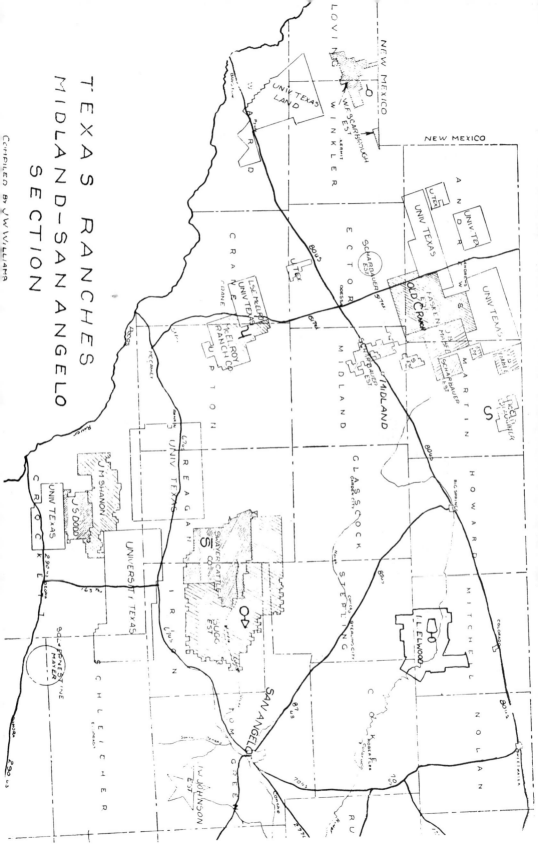

County in a rich, beautiful section of the state. But the powers that made laws did not remain constant in their original policy— and the outcome was what seemed to be a pretty bad break for the University of Texas. A lot of land was substituted for the original million beautiful and fertile acres. Most of the new gift was stretched out along the valley of the Pecos River and in the South Plains country where rainfall was scant and vegetation was none too plentiful. Part of it was covered with sand dunes higher than a two-story house. Some of it was in the extreme western end of the state where the rainfall amounted to only ten inches per year.

But the guardian angel or higher education could see deeper in the ground and farther into the future than those who made the laws. A million and a half acres of this endowment land in sixteen tracts were scattered through what oil men now know as the Permian Basin. No geologist could have selected these University lands half as well. If a giant patch 200 miles long and 100 miles wide could be placed on West Texas with its length extending northwest and southeast up the Pecos River and with its center about thirty-five miles south of Odessa, this patch would just about cover all sixteen of these tracts of University land.[47] The same patch would cover a great part of the oil production in the Permian Basin. Up to now seven of these sixteen tracts are producing oil. One hundred and seventy-seven million dollars chiefly from oil royalties, lease bonuses and rentals[48] have been collected since 1923 when the first well was brought in on University of Texas land ninety miles west of San Angelo. Before that date, these 1,500,000 acres (together with another 500,000 acres in the El Paso country) had produced no revenue except from grass. The University had not been in the cattle business directly; these lands were leased to cattlemen and the income from the leases applied to current expenses. The policy has continued through the years, but the income from grass has become relatively much less important since the discovery of oil thirty-one years ago. The University income has been similar to that of any other big rancher who has struck oil. The University of Texas, with some-

47. Any good map of each of the various counties in which University holdings are located will give an accurate picture of these large bodies of West Texas land. The maps referred to in particular were: the Texas Land Office maps of the following counties: Andrews 1925, Crane 1902, Crockett 1901, Dawson 1900, Culberson 1908, Ector 1901, Gaines 1922, Hudspeth 1917, Irion 1893, Loving 1902, Martin 1900, Pecos 1917, Reagan 1915, Schleicher 1898, Terrill 1905, Upton 1918, Ward 1902, Winkler 1901.

48. Texas Almanac, 1954-1955, 382. This fund has more than doubled since 1948.

what in excess of 2,000,000 acres,[49] is the greatest land owner in the state. Other than legislative appropriations, oil and cattle are the two chief sources of its income.

My jouney is shortly to enter the country in which the second largest land ownership in Texas is located.

I stopped and ate breakfast at a cafe on Odessa's long main street. The town had experienced a marvelous growth since my last visit there a few years ago. It is now pretty near the center of the far flung oil development in the Permian Basin. My first glimpse of Odessa was more than forty years ago. As I remember now, it then had one small frame hotel and a few business houses, but it was heart and soul a cow town. Even now it has not lost its western cow country atmosphere. Much the same comment may be made about Midland, Odessa's twin city, twenty miles to the east—except that Midland grew first and has had all through the years a greater concentration of ranch owners than has Odessa. The Cowdens and a number of other old families are still strongly intrenched in Midland's financial make-up.

After breakfast I headed west toward the immense trans-Pecos part of Texas—a country so vast that it seems too large to be perched under a single sky. Perhaps it is fitting that the largest private land ownership in Texas (second only to that of the University of Texas) should be located out there. Surely it will be a surprise to most readers to know that neither the King Ranch nor the former Matador Land and Cattle Company can claim to be the largest private land owner. One giant spread of acreage scattred across more than a dozen counties is larger than the acreage totals of both of those great ranches—so great that the sun rises on the portion of this present day empire that lies near San Angelo twenty-four minutes earlier than upon the part near El Paso.

Its story is a hundred years old. It was 1849 when the California Gold Rush was new that more than one wagon road was made westward across Texas through what is now El Paso county. Captain Randolph B. Marcy of the United States Army made one of these roads—one that reached from El Paso County to northeast Texas. His old wagon road had much influence on subsequent

49. Ibid, 352. The exact acreage is 2,329,168.

history.[50] In 1858 when the United States government instituted its first overland mail service to the Pacific Coast, the Postal Department ran the new route parallel to Marcy's road and actually made use of more than one hundred miles of his old wagon tracks. The attention of railroads was focused on the same route. Two of these, the Memphis, El Paso, and Pacific and the old Southern Pacific (not the present Southern Pacific) intended to build westward from northeast Texas, began before the Civil War.

Then came the Texas and Pacific chartered by the United States Government in 1871.[51] By 1880, under Jay Gould it took over both of these early roads and began the now famous race with the present day Southern Pacific, then under the direction of C. P. Huntington. Huntington, who was building the chief branch of his road from the west, beat Gould, who was building from the east, to El Paso and ran this road through the mountain passes as far as Sierra Blanca ahead of Gould. Lawsuits threatened to hamper the operations of the two roads, but these two financial giants sat down and compromised their differences. They divided their railroad world between them. The details of the agreement are not important here except that Gould was given a permanent right to use the Huntington tracks into El Paso.

But of special import to us are some of the facts about Jay Gould's railroad. His road, the Texas and Pacific, had service extending completely across the state. It had built more than 900 miles of track which the state of Texas had obligated itself to reward with a land grant of from sixteen to twenty sections for each mile. Altogether, the state thus barring technicalities owed the Texas and Pacific railroad 12,723,200 acres of land, but so many large obligations of this type accrued so suddenly that the public lands were exhausted and only about 5,000,000 acres[52] of this great land grant was ever delivered. Nevertheless, it constituted the greatest amount of land ever transferred by the state

50. In 1854, Brevet Captain John Pope (later General Pope) at the expense of the United States Government, made a survey for a proposed railroad that followed Marcy's wagon road at least half way across Texas and closely paralleled the remainder of the road. See **Reports of Explorations and Surveys to Ascertain the most Practicable and Economical Route for a Railroad from the Mississippi River to the Pacific Ocean**, Vol. 2, Sen Ex doc. No. 70, 33 cong. 2 sess. Pope's report, pp. 1-111, gives details of route. The road when finally built by the Texas and Pacific Railway Co. did not follow Marcy's wagon road but it did parellel it.

51. For a good discussion of this railroad and its lands see C. S. Reed, **A History of the Texas Railroads**, 356-375, 544-551.

52. There were various reasons why the Texas and Pacific was not awarded the full amount as promised. It actually received 5,167,120 acres and in a law suit lost 246,056 acres back to the State. See S. G. Reed, **A History of the Texas Railroads**, 371.

of Texas to one private establishment. The grant was about equal to the area of the State of Massachusetts.

Other railroads received large land bonuses, but in the main their grants were widely scattered. This Texas and Pacific land grant of 5,000,000 acres was largely located in a long strip of land reaching across West Texas that was partly sixteen miles wide and partly eighty miles wide. Midland and Odessa are both within this strip of land, and the Texas and Pacific railway itself is built along its course. In due time, as required by law, the Texas and Pacific railroad disposed of all of this bonus land; but as it happened, Fraser, Burr and McAlpin have acquired a large remnant of it and until this day still own more than 1,700,000 acres.[53] They are the largest private land owners in Texas. They own no such large solid blocks as does the University of Texas. Their lands are somewhat scattered in Sterling and Mitchell counties northwest of San Angelo. But they own 31,000 acres in these counties alone. Somewhat similar are their holdings in Glasscock, Reagan and Upton counties, but farther west, in the country that is too dry for farming except by irrigation, their lands usually assume a more definite pattern.

In accord with the general rule, the land grants to the Texas and Pacific railway company required that the company itself bear the cost of surveying double the amount of land awarded it by the state. Only the alternate sections (square miles) then became the property of the railway company while the other half of the land blocks thus surveyed became a part of the endowment of the Texas Public School system.

During the years from about 1874 to 1880 the Texas and Pacific surveyed a body of land nearly 400 miles long reaching eastward from El Paso, with several smaller blocks scattered across the counties as far as the west bank of the Brazos River within about fifty miles of Fort Worth. They had blocked out on the map of West Texas an area that looked like a giant checkerboard twice as large as the state of Massachusetts. Only half of this great tract was theirs, but it did constitute an immense body of land.

53. Fraser, Burr and McAlpin (recently changed to Frazer, Burr and Olyphant) are credited by the tax rolls (for 1947 or 1948) of the various counties as owners of the following amounts of land; Sterling County, 21,298.66 acres; Mitchell, 10,803 acres; Glasscock 49,958; Reagan 17,957.7 acres; Upton 44,541.9 acres; Loving 89,843.58 acres; Reeves 256,738.58 acres; Pecos 65,706.58 acres; Jeff Davis 96,436.69 acres; Presidio 42,246.69 acres; Culberson 523,061.5 acres, Hudspeth 384,862.44 acres; and El Paso County 121,158.76 acres—a grand total of 1,724,624.08 acres. This does not include smaller amounts in Ector and other counties.

Frazier, Burr and McAlpin, who own the large remnant of this immense land grant, are not in the cattle business and, so far as I know, are not in the oil business. They lease their land both to cattlemen and to oil companies and depend upon the income from these sources for their revenue. They own only alternate sections as did the Texas and Pacific Railway before them. Several rather large ranchers have acquired a number of the alternate school sections in between the Frazier, Burr and McAlpin lands and have squared out their pastures into solid blocks of grazing land by leasing the property of these successors to the Texas and Pacific.

This policy of the State of Texas which required railway companies to own only alternate sections has caused no end of trouble to many ranchers. If, as may be the case, the practice was justified on the broader grounds of the public good, it has at least set up a problem for every ranchman who acquired a block of railway lands. He, in every case, sooner or later had to do something about the half of his block of land which he did not own. Some ranchers traded sections on the outside of their block for sections acquired by nesters or others within it. Some of them purchased the smaller holdings outright. Some of them were fortunate enough to obtain grass leases on these school sections under arrangements that were not subject to abrupt change. But in the main, it took years to work out the problem so as to eliminate uncertainty. Sometimes, as we shall see later, a rancher lost his fortune trying to solve this problem.

I continued my drive westward from Odessa. Twelve miles from town, I came to the west edge of the High Plains. Here the surface of the earth drops down something more than one hundred feet within a mile or two. Here and for twenty-five miles to the west, the land is extremely sandy. Toward the west part of that sandy belt are the famous Sand Hills of West Texas. Many of the sand dunes of these hills are made of pure sand with no vegetation growing upon them; some of them appear to be from twenty-five to fifty feet high, but many of them are not so prominent, and there is quite a little growth such as shinnery between dunes and even well up the sides of some of the dunes themselves.

This sandy belt is made up chiefly of lands that were set aside for the endowment of the public schools and the University of Texas. The Texas and Pacific surveyors skipped this belt and also skipped another twenty-five miles to the west of it. They began to again select lands in Loving County, some twenty miles north of the town of Pecos. Frazier, Burr and McAlpin still own nearly

90,000 acres of these original surveys in Loving County.

But the big holdings of former Texas and Pacific lands are west of the Pecos River. Reeves county, the first of these, joins the river. Frazier, Burr and McAlpin hold 256,000 acres in this one county. Next is Culberson County in which they hold 523,000 acres; next Hudspeth County with 384,000 acres; and finally comes El Paso County in which these gentlemen hold 121,000 acres. They have smaller holdings in Pecos, Jeff Davis and Presidio counties which amount to 65,000 acres, 96,000 and 42,000 respectively.

The four counties from the Pecos River to El Paso in which Frazier, Burr and McAlpin hold their largest acreage are in general not the site of Texas' greatest ranches. Perhaps this is due to the presence of so much land which may be leased. J. B. Foster, Velma C. Rounsaville, Gresham and Hunter, and the West Cattle Company own ranchlands (chiefly in Culberson County) that vary from above 50,000 to well above 100,000 acres.[54] Let us make a more complete reference to the West Cattle Company at a more appropriate place.

Another of the large ranches that holds lands in these four counties is the Reynolds Cattle Company. I drove westward and on toward the south—through the town of Pecos and on to Fort Davis partly to view this big ranch. The first forty miles of my drive from Pecos was across a dry country where farming would be next to impossible except by the aid of irrigation. Even ranching would appear to be a hazardous business there. But shortly past Balmorhea, I entered the Davis Mountains and evidence of drouth began to be less prominent. It had just stopped raining when I arrived at Fort Davis in the heart of the mountains; water lay everywhere in the streets of this small western town.

The altitude is over 5,000 feet—high and cool enough to attract summer tourists and blessed with enough rainfall to make it a great ranching country. A hundred-mile scenic drive loops

54. In addition to these, according to 1948 tax rolls W. D. Johnson has 58,625.09 acres in Reeves County, and Popham Land and Cattle Co. in both Reeves and Pecos counties has 63,561 acres. E. C. Bolton and associates have about 160 sections of land in Culberson County; Velma C. Rounsaville in Reeves and Culberson counties has 100,966 acres; and J. B. Foster has 108,451.73 acres in Culberson. Gresham-Hunter Corporation has 20,673 acres in Reeves County and 101 sections (of 640 acres each) in Culberson County while J. C. Hunter has more than 50 sections in the latter county. The O'Keefe Estate has 72,449 acres in Hudspeth County; the Pierces and Dempsey Jones of Ozona, Texas also have 99 sections in the county. In the same county are J. Q. Adams with 67,136.4 acres, Guitar trust with 51,488 acres. Ray Willoughby of San Angelo has more than 50,000 acres in Culberson and Hudspeth counties. A number of other ranches in these same counties, large because they include leased lands, are not included in this account.

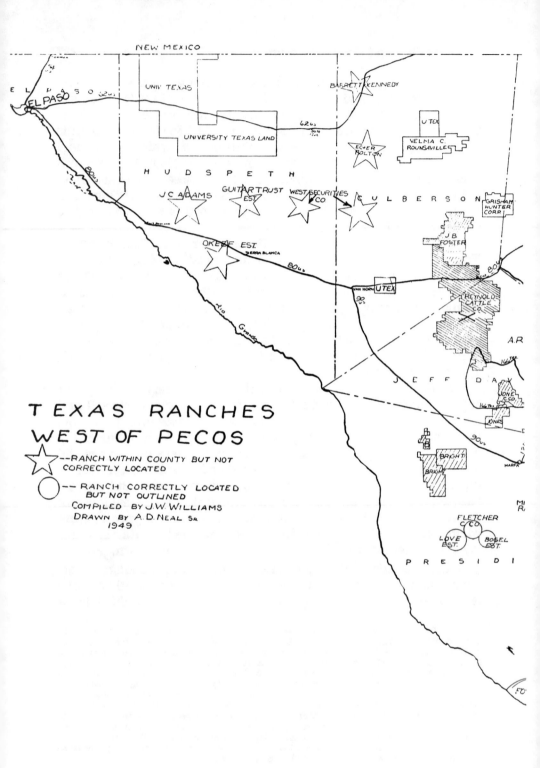

TEXAS RANCHES
WEST OF PECOS

☆ --RANCH WITHIN COUNTY BUT NOT
CORRECTLY LOCATED

○ -- RANCH CORRECTLY LOCATED
BUT NOT OUTLINED
COMPILED BY J.W. WILLIAMS
DRAWN BY A.D. NEAL SR.
1949

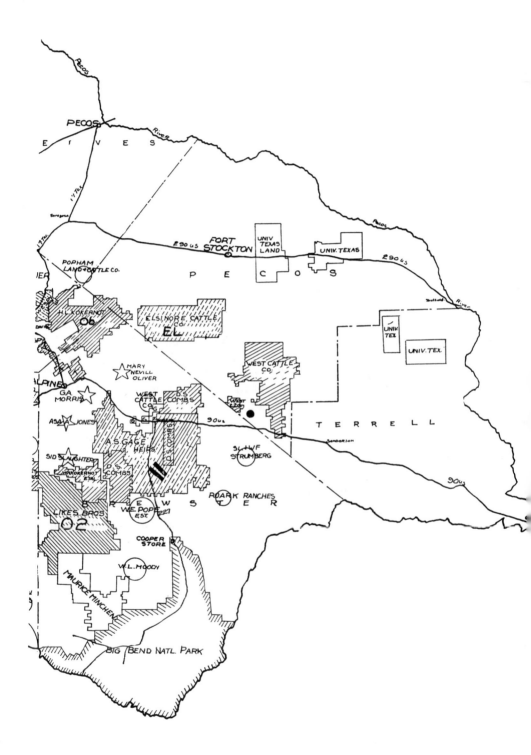

westward through the mountains and back into Fort Davis. Somewhat as one who seeks adventure, I turned my course from Fort Davis still deeper into the mountains along this interesting road. About forty miles from Fort Davis among granite mountains and green meadows, this beautiful driveway enters the southern part of the Reynolds Cattle Company's largest ranch. Their pasture here is between fifteen and twenty miles wide but it was more than forty miles at its greatest length, a dimension that runs nearly north and south. They reach northward past the town of Kent on the Texas and Pacific Railway; and here at the south end they extend to spectacular Sawtooth Mountain, which like a sawtooth shaped tombstone stands several thousand feet high directly in front of my car. The Reynolds were among the first ranchers to enter West Texas. For a number of years of frontier history their activities centered around old Fort Griffin and the town of Albany. As early as 1877 their ranch covered the east part of Haskell County.[55] The springs now included within a park in the town of Haskell was one of their watering places. In 1882 they organized the Reynolds Cattle Company[56] which has since spread to the Texas Panhandle and to the Davis Mountains. The present big ranch out here among these mountains covers 180,000 acres[57] owned by them in fee and many thousand of acres of the Frazier, Burr and McAlpin lands.

I drove on around the west side of Sawtooth Mountain, on past Mount Livermore, and back toward the east, still following this hundred-mile mountain drive. But I did not return to Fort Davis. It was more in keeping with my plans to turn southward to Marfa, in Presidio County, one of the large West Texas Counties that extends into the Big Bend country and borders on old Mexico. Strangely this large county with much color in its early ranching history is not the center of the great ranches[58] of this extreme

55. Interview with the late Bud Matthews of Albany, Texas.

56. Gus L. Ford, Texas Cattle Brands, 136.

57. According to the 1947 tax rolls, Reynolds Cattle Co. has 67,159.49 acres in Culberson County and 113,023.33 acres in Jeff Davis County.

58. A hasty check of Presidio County (1947) tax rolls shows C. T. Mitchell & Company with 32,530 acres, W. K. Mitchell with 8,320 acres, T. Clay Mitchell, Jr. with 14,571 acres (and T. Clay Mitchell, Jr. Walker Ranch with 3,397 acres), and W. B. Mitchell's Sons with 16,530 acres. Fowlkes Brothers (1948 tax rolls) have 167,025 acres —possibly including some duplicate items. Mrs. L. C. Brite had (1947 rolls) 61,730 acres; Hester Brite Dunkle 37,920 acres; Nancy Dunkle Roderick and Jane Dunkle White had 27,680 acres. Lykes Brothers (1947 rolls) had 32,625.95 acres in Presidio County which is only a small extension of their large ranch in Brewster County. The heirs to the great Gage Estate, Roxana Gage Catto and Dorothy Gage Forker have 18,025 and 10,660 acres respectively in Presidio County. Their chief holdings are in Brewster County.

western portion of Texas. Fowlkes Brothers own the largest spread in the county with their 165,000 acres owned in fee and probably quite a leased area beside. Their ranch lies in the south part of the county against the Rio Grande with old Mexico for a next door neighbor. Joining them on the northeast is Rawls Brothers who own about 60,000 acres, and some distance northwest of them (30 miles southwest of Marfa) is the equally large W. E. Love Estate. Nearly twenty-five miles west of Marfa is the somewhat larger Bright Ranch, the scene of a rather serious gun battle that originated across the border during the Villa- Carranza War in Old Mexico. The several Mitchell ranches, with a combined area of more than 100 sections, are interesting because of the founder, W. B. Mitchell who pioneered this mountain country in 1882. But there is another character who deserves special mention among the early settlers of Presidio County. Milton Favor,[59] first of its big ranchers has become a legend of the land. He came from eastern United States about the year 1850 and soon settled in the Big Bend some forty-five miles southwest of this present-day town of Marfa. With his Mexican wife he seems to have learned rapidly how to adapt himself to the ways of his neighbors from across the border. He constructed Fort Cibola (the ruins of which still stand southwest of Marfa) and soon built up a large ranch with his long-horned Mexican cattle. In spite of Indian raids his herds are said to have reached the large total of 15,000 head by about 1880. Almost as famous as the F brand which his cattle wore, was his peach brandy known far and wide throughout the Davis Mountains and Big Bend Country. According to a local tradition he died a pauper largely because of cattle thieves.

But we must move on from Marfa over the twenty-six mountain miles that lead to Alpine, the real center of the big ranches that lie west of the Pecos River. Ranches that are large by usual standards[60] are that of J. W. Espy and sons with 85,000 acres located both north and south of Fort Davis; the W. E. Pope

59. J. Evetts Haley, **Jeff Milton A Good Man with a Gun** (Norman, Oklahoma— University of Oklahoma Press, 1948), 74-76; Carlyle Graham Raht, **The Romance of Davis Mountains and Big Bend Country**, 225.

60. From the 1947 tax rolls of Jeff Davis and Brewster counties; Jones Cattle Company has 40,040 acres; Frank Jones 9802 acres; W. T. Jones 7618 acres; H. D. & W. F. Cornell 44,579 acres; Coffield & Gearhart 42,256 acres; W. E. Pope Estate 116, 750 acres; Maurice Minchen 136, 986 acres; Combs Cattle Company 57,131; David S. Combs Trust 30,620, David S. & W. F. Combs 14,118 acres; Guy S. Combs 17,183; David S. Combs Jeffers 14,920; Asa A. Jones 50,532; Ernest Jones 47,428; Mrs. Martha Oliver and others 60,768; C. A. Morris 40,125; I. C. Roark 73,874; F. M. Roark 12,060; J. C. Roark 3,840; E. L. Strumberg and wife 47,102; S. L. Strumberg and Hinson Davis 12,735; H. G. Towle 38,721, Victor Pierce 12,488; W. B. Mitchell's Estate 9,929; W. L. Moody, III, 24,960.

Estate with 116,000 acres near the center of Brewster County southeast of Alpine; the ranches of David S. Combs and Combs Cattle Company with a combined area above 130,000 acres about thirty miles east and an equal distance southeast of Alpine; and that of Maurice Minchen who owns (but does not operate) 135,000 acres in the Christmas Mountains far to the south of Alpine joining the Big Bend National Park. Incidentally this great park of 707,000 acres that fits snugly down into the tip of the Big Bend (of the Rio Grande) of trans-Pecos, Texas, is, of course, not a ranch but it is grazing land.

There are no Texas ranches west of the Pecos River equal to this large park and no land ownerships as large except the University of Texas and Frazier, Burr and McAlpin. But there are an exact half dozen ranches that may be classed among the great or the near great of the entire state of Texas. All of them are at their nearest borders within fifty air-line miles of Alpine. One of the six, the Reynolds Cattle Company far to the northwest of this town, has already been mentioned.

Almost at the city gates of Alpine and extending thirty miles to the north, are the lands of H. L. Kokernot and Son that make up another of these large ranches. Watered by 500 springs and covering 217,000 acres[61] of mountain and meadow this old estate at the east edge of the Davis Mountains is one of the classics of western grazing lands. The 06 brand which for a long time has been the symbol of a Kokernot cow was purchased by L. M. Kokernot in 1873. Later it became the property of the late H. L. Kokernot who has been a prominent rancher of south and west Texas for more than fifty years. W. H. Kokernot and other relatives own 87,000 acres[62] thirty miles south of Alpine, bringing the total Kokernot holdings in the Alpine country up to 304,000 acres.

Lykes Brothers, who own a steamship line at Houston, Texas, during the 1930's purchased the 02 Ranch built up by W. W. Turney, a pioneer of Brewster County. Turney had bought the 02 brand from Gage Brothers in 1891 and within a little more than a dozen years afterward had expanded his holdings into a

61. In Jeff Davis County 143,037.78 acres (1948 tax rolls), Pecos County 17,868.80 acres (1948 tax rolls) and in Brewster County 56,249.32 acres (1947 tax rolls)—total 217,155.90 acres.

62. The 1947 Brewster County tax rolls show Miss Josephone Kokernot with 23,168.5 acres, J. W. and L. G. Kokernot with 12,589.08 acres, L. M. Kokernot with 17,002 acres, Mrs. Virginia Kokernot with 19,387.6 acres and W. H. Kokernot with 15,372.1 acres. The total is 87,519.28 acres. The total for the various Kokernot holdings in Jeff Davis, Pecos and Brewster counties is 304,675.18 acres.

large ranch.[63] Turney's lands blocked midway along the west line of Brewster County (and reaching over into Presidio County), had come from the Galveston, Houston and Henderson and the Texas and St. Louis railway companies. As was true in general, these railway companies owned only the alternate sections and Turney assumed the momentous burden of purchasing the remaining lands within his pasture. Unfortunately it proved too great an undertaking during the years of shrinking markets just ahead of him. He failed in business and lost his ranch largely because of this effort to cement the property into a solid block.[64] Lykes Brothers now hold this great spread of 265,000 acres[65] of ranch lands which is another of the very large ranches of western Texas.

E. L. Gage, another pioneer cattleman of this part of the West, came to the Big Bend in 1880 and opened a ranch at McKinney Springs, south of Marathon.[66] Both E. L. and A. S. Gage were established in the cattle business of the area before 1890. But it was sometime after 1900 when A. S. Gage emerged as one of the big cattlemen of his area.[67] Now his two heirs, his daughters Mrs. Roxana Gage Catto and Mrs. Dorothy Gage Forker, own 195,000 and 203,000 acres[68] respectively of Brewster and Presidio County lands. The total of 398,000 acres was for some years operated as a single ranch.

After a short stay at Alpine, I drove toward Fort Stockton which is sixty-two miles to the northeast. At a little more than half the distance, I began to cross the 186,000 acre pasture[69] of the Elsinore Cattle Company. Giddings and Lennox the present owners had acquired the property from J. S. Lockwood in 1900, after he had operated the early ranch since 1886.

63. Gus L. Ford, **Texas Cattle Brands,** 135.

64. The **Wichita Falls Record** News of August 7, 1936 published the Associated Press story of the forced sale of the W. W. Turney Ranch. The debt that had to be satisfied had been made to a great extent by the purchase of many squatter sections from 1905 to 1907. W. W. Turney was the first county attorney of Brewster County. He was elected in 1887. See Carlyle Graham Raht, **The Romance of Davis Mountains and Big Bend Country,** 301.

65. The 1947 Brewster County tax rolls show 232,606.6 acres (subject to a possible error in addition). This amount added to the 31,948.35 acres of the ranch in Presidio County give a total of 264,554.95 acres.

66. Carlyle Graham Raht, **The Romance of Davis Mountains and Big Bend Country,** 224.

67. Gus L. Ford, **Texas Cattle Brands,** 128.

68. The 1947 Brewster County tax rolls show Mrs. Roxana Gage Catto with 176,639.29 acres, and Mrs. Dorothy Gage Forker with 192,987.64 acres. Their combined holdings in both Presidio and Brewster Counties are 194,665.17 acres and 203,648.45 acres respectively or a grand total for both of 398,313.63 acres.

69. The 1948 tax rolls of Pecos County show that the Elsinore Cattle Company had 186,394.67 acres.

Upon arriving at Fort Stockton, I completed—or nearly completed—my information on the West Cattle Company, the largest ranch west of the Pecos River. From the records of several counties, I had already learned that West securities company held 54,000 acres in Hudspeth and 112,000 acres in Culberson County; that J. M. West held 47,000 acres in Brewster County; and that the West Cattle Company held 8,000 acres, also in Brewster County. Now I discovered that their principal ranch covered 199,000 acres that lay thirty miles south of Fort Stockton and extended far down into the point that made up the south corner of Pecos County. I later learned that the West Production Company owned some 32,000 acres in Dimmit County and that J. M. West et al owned 60,000 acres in Maverick County, making a grand total of 502,000 acres[70] which I have presumed to be a single ranch.

It was not more than ten o'clock at night when, because of an early drive planned for the next morning. I went to bed in Fort Stockton. The intended drive was to take me to Ozona, county seat of Crockett County, which joined Pecos County on the east and to get me there by eight o'clock when the tax collector's records would be available. Why the early drive when the journey was only the distance from one county seat to the next? Most court houses in adjoining Texas counties are only about thirty miles apart—and the three (in Fort Davis, Marfa and Alpine) which I had just visited were nearer together than that. The truth is that the proximity of these last three county seats, situated in such large counties as they are, is something of a geographical accident. But no such accidents had played in my favor between Fort Stockton and Ozona. The distance was 112 miles!

I was fourteen miles on my way at sunrise. Here in the middle of a large block of University of Texas lands the road to San Angelo turned away from my highway sharply to my left. After driving twenty-two miles from this point, I was surrounded with oil wells at the village of Bakersfield in the heart of another large tract of University lands. An additional twenty miles placed me opposite the great Yates oil field which lay only a few miles

70. The West Securities Company has 112,396 acres in Culberson County and 53,705.1 acres in Hudspeth County (1947 tax rolls); the West Cattle Company has 7,695 acres in Brewster County (1947 tax rolls) and 198,752.18 acres in Pecos County (1948 tax rolls); J. M. West has 47,489.62 acres in Brewster County (1947 tax rolls) and J. M. West et al have 59,909 acres in Maverick County (1948 tax rolls); and finally the West Production Company has 32,253 acres in Dimmit County (1948 tax rolls). The grand total is 502,199.2 acres.
The Allison Ranch Company of Fort Stockton has (1948 tax rolls) 67,237.49 acres.

to the left between my highway and the Pecos River. Fifteen miles ahead I had breakfast at Sheffield—but had not yet driven far enough to get out of Pecos County.

My road soon crossed the Pecos River and wound around the edge of a cliff which gave a breath taking view of chalk-white bluffs and rugged canyon scenery, a color picture which would have made a magazine cover pleasing to the most discriminating eye. But likely as not the publicity men will soon find the place, for the pattern of the great Permian Basin oil area seems unfinished in the direction of Crockett County, and both the multitudes and the publicity men will in due time follow if there are big oil discoveries.

Meanwhile these immense if picturesque solitudes, the scarcity of people and human habitations along the highway, and the vast mileage between towns cause Crockett County to seem like a little empire of its own. Ozona, the county seat and only town within the county boundaries, had all the appearance of the capital of such an empire. The beautiful homes and substantial looking business establishments gave the place a self-sufficient look that is rare even in the great wide open spaces of Western Texas.

It was a little after eight o'clock when I walked into the tax collector's office of this small town and found one of the best kept tax rolls which I had yet encountered. In record time I was able to find from these well kept rolls the 53,000 acre J. S. Todd Estate northwest of town and the 132,000 acre J. M. Shannon Estate[71] which joins Todd on the north and to find that other ranches, though a number of them were large and independent estates, were not quite the size required here. The late J. M. Shannon was not a stranger to some of the other ranching areas already visited; he built the fence around much of the southern part of the old XIT Ranch[72] during the middle 1880's.

The University of Texas has two large tracts of land in Crockett County, one eight miles west of town (which I had driven through on the way to Ozona) and the other fourteen miles north of town. Two other tracts of University land extend southward into Crockett out of Reagan and Upton Counties.

My trek through the ranch country extended east from Ozona to Sonora, north from that town to San Angelo, and westward from there to the little town of Mertzon in Irion County.

---

71. From the 1947 tax rolls, the J. S. Todd Estate owned 52,745.95 acres and the J. M. Shannon Estate, 131,654.7 acres.
72. J. Evetts Haley, **The XIT Ranch of Texas**, 92.

The Sol Mayer Ranch of 70,000 acres[73] lay to the north of my road, mid-way between Ozona and Sonora, and the J. W. Johnson estate of 85,000 acres[74] lay to the east of my road in Tom Green County. Most of it was in the Christoval country southeast of San Angelo. But neither of these approached the magnitude of some large ranches found west of San Angelo.

Nearly fifty miles west of that important cattle and sheep town was the 172,000 acre spread belonging to the Sawyer Cattle Company. Their lands located in the valley of the middle Concho River stretched across the line between Irion and Upton counties with nearly two thirds of their acreage[75] in the latter county. Even before 1880 Sterret and Sherwood brought the Bar S brand of cattle to this part of the middle Concho Valley. In 1884 they sold to Sawyer, McCoy and Rumry who organized their holdings into the Sawyer Cattle Company.[76] Their large cow outfit still holds its spot in the Concho Valley at the end of seven full decades of history.

Joining the Sawyer interests on the east and north is a somewhat larger ranch that should cause us to feel that we are paying a visit to an old neighbor. In the 1880's when Dan Waggoner began to throw his great herds into Indian Territory across Red River northwest of Wichita Falls and when Burk Burnett began to occupy the Indian country east of Waggoner, E. C. Sugg and his brothers took up a similar position east of Burnett.[77] The Suggs ranched in the Indian country northeast of Wichita Falls— the country that lay across Red River from Clay County, Texas. They enjoyed a great period of expansion here until shortly after 1900, when a change in government policy made it necessary for them to seek other grazing lands. While across Red River they purchased a large herd of cattle from Wilson and Davidson of Gainesville, Texas. The herd was branded with the now famous O. H. Triangle which the Sugg brothers adopted and kept.[78] Upon leaving the Indian Territory, they moved to this position joining the Sawyer Cattle Company on the middle Concho. Altogether

73. Sol Mayer and Ernestine trustee according to 1947 tax rolls had 42,115 acres in Sutton County and 23,710 acres in Schleicher County. Sol Mayer individually was credited with 3,589 acres in Schleicher.

74. The J. W. Johnson Estate held (1947 tax rolls) 85,653.08 acres in Tom Green County.

75. The Sawyer Cattle Company (1947 tax rolls) held 60,232 acres in Irion County and 112,260.4 acres (1948 tax rolls) joining it in Reagan County—total 172,492.4 acres.

76. Gus L. Ford, **Texas Cattle Brands**, 127-128.

77. Interview with the late Bob McFall of Wichita Falls, Texas.

78. Gus L. Ford, **Texas Cattle Brands**, 61.

they acquired the 275,000 acres that is still owned by the Sugg heirs.[79] Their great pastures include many square miles of land owned by Fraser, Burr and McAlpin and others; but this is no longer a serious handicap to the owners, since they now lease the holdings to other ranchers instead of operating the property themselves.

In the valley of the Colorado River northward and slightly toward the east from these Sugg lands, across a gap of some twenty-five miles in the greatest of the ranches that make up the W. L. Ellwood Estate and perhaps the last large reminder in the cattle country of the early day fortunes made in the barbed wire industry. This pasture alone covers 121,000 acres. With the lands of Ellwood and Towle that join it on the west and the property of Miss Patience Ellwood along the south rim of the ranch it totals 150,000 acres. With the addition of the remnant of their big ranch in Hockley County (west of Lubbock) and 126,000 acres recently acquired in New Mexico the Ellwoods still hold 298,000 acres.[80] Once they owned about 400,000 acres—that was before the coming of the Santa Fe Railroad and the advance of the plow in Hockley and Lamb Counties. They still burn the old Spade brand into the hides of their cattle. So do Ellwood and Towle with only a slight variation in the way the brand is applied.

This coverage of the Ellwood lands completed the big pastures of the San Angelo country. It was now time to turn my car around and start toward home while I still had a clean shirt left in the bag. But I stopped on my way at the office of the Swenson Land and Cattle Company in Stamford, Texas.

W. G. Swenson, their manager, very generously opened his files in complete cooperation with my search for information.

79. The 1948 tax rolls show that Mrs. E. C. Sugg owns 4,739 acres and Calvin H. Sugg 11,763.36 acres in Sterling County and that Calvin H. Sugg owns 46,319.75 acres in Reagan County. The 1947 tax rolls show that Mrs. E. C. Sugg owns 79,291.1 acres, Calvin Sugg owns 30,310.1 acres and the A. A. Sugg Estate owns 63,869 acres in Irion County—also that A. A. Sugg, Jr. and Sammie S. Farmer own 25,215 acres in the same county. Also the 1947 tax records show that Calvin H. Sugg owns more than 14,000 acres in Tom Green County. The total is about 275,500 acres although the figures from Tom Green County are not exact. However, the lands belonging to the Sugg heirs have been thus divided and leased out to other operators and hence are no longer operated as one large ranch.

80. Without the lands of Ellwood and Towle of 19,016.6 acres (1948 Mitchell County tax rolls) and the 10,710.45 belonging to Mrs. Patience Ellwood in Sterling County (1948 tax rolls) the Ellwood Estate has 142,649.87 acres left (according to 1948 tax rolls) in four Texas counties. These items by counties are as follows: in Hockley County 21,754.02 acres, in Mitchell County 88,320 acres, in Sterling County, 20,215.65 acres and 12,360.2 acres in Coke County. To this total for the Ellwood Estate add 126,000 acres in New Mexico (reported in The Cattleman, August, 1947) and the grand total is 268,649.87 acres.

The largest of the Swenson ranches covers most of the northwest quarter of Throckmorton County and amounts to 106,000 acres. Next in size the Tongue River Ranch west of Paducah covers 79,000 acres, next the remnants of the old Spur Ranch nearly 71,000, and smallest of them all the Flat Top Ranch just west of Stamford has 42,000 acres. Altogether these Swenson properties amount to just less than 298,000 acres.[81]

S. M. Swenson, great uncle of the present manager, came to Baltimore from Sweden and later came to Texas while it was still a republic.[82] He became a merchant in Richmond, then a shipping point on the Brazos River, and later became both a merchant and hotel proprietor in the new capital city of Austin. He must have prospered in his business, for when it came time for the Buffalo Bayou Brazos and Colorado Railroad (first in Texas) to market its premium lands, he bought nearly 100,000 acres of the issue. This occurred before the Civil War. Later he purchased the school sections which alternated with these railroad lands and thus cemented his holdings into solid blocks. These are the very same bodies of land that now constitute the Throckmorton County and Flat Top Ranches of Swenson Land and Cattle Company. Also the former Ericsdale Ranch southeast of Stamford (since disposed of) was a part of the land acquired from this early day Texas railroad.

Ranching on these large blocks of land came as an afterthought. Since the land was subject to taxation, the elder Swenson in the early 1880's decided to put his two sons Swen A. and Eric P. in the cattle business in order to derive some income from his investment. The boys adopted S. M. S., the initials of their father, as their brand. About this same time A. J. Swenson, nephew of the Austin merchant, came over from Sweden and soon went to work on the Ericsdale Ranch. Shortly he sought other work, but in 1897 returned to the employ of Swenson Brothers as bookkeeper. After twenty years or more, he became manager, a position which he held until his recent retirement. W. G. Swenson, the present manager is his son. As the management of the

---

81. From the files maintained by the owners of the Swenson Land and Cattle Company at Stamford, Texas, the following information was obtained: Their Throckmorton County Ranch covers 106,047 acres; their Tongue River Ranch at the corner of Motley, Cottle, King and Dickens counties covers 79,165 acres; the remnants of the old Spur Ranch in Dickens, Garza and Kent counties cover 70,606 acres; and their Flat Top Ranch, chiefly in Stonewall County covers 42,030 acres. The total is 297,848 acres.

82. For a historic sketch of the Swenson Ranches see Mary Whatley Clarke, "SMS Ranch", The Cattleman, March 1948, pp. 78, 80, 82. Supplementary facts have been obtained by interviews with the late A. J. Swenson and with W. G. Swenson.

property has been handed down, so has the ownership been passed along to the heirs of the pioneer Swedish merchant who founded the ranch. His grandson, S. M. Swenson of New York City, is now president of the greatly expanded business. The SMS brand is now one of the well known symbols of the cattle world.

In about 1900 when it became known that the Texas Central Railroad would be extended to the site that is now Stamford, the Swenson Brothers decided to sell the Ericsdale Ranch, since the new railroad was to build across the corner of their land. They subdivided the land into farms and offered them for sale. At the suggestion of A. J. Swenson, Swedish people of Lutheran faith were invited to form a colony on the ranch. The plan was endorsed by Pastor Stamline, head of the Lutheran Church in Texas, and the Swensons offered to give eighty acres of land within the proposed community for the establishment of a church and to pay $300 per year for three years on the pastor's salary. The colony was organized, and the little church was built as planned. The church house now in use is a nice brick structure which stands near the banks of California Creek ten miles southeast of Stamford. Oil has enriched the Swedish community and has even been discovered on the eighty-acre church block. The Swensons who live in Stamford continue to attend the little church out there in the country among the children and grand-children of the Swedish farmers who moved out West when the Texas Central Railroad built across the corner of the Ericsdale Ranch.

I returned to my home in Wichita Falls but did not make the long loop through the ranch of South Texas as had been planned. Perhaps it is just as well that the reader be spared the details, since the chief purpose of this chapter is to list and as far as possible, rank the dozen or more greatest ranches in Texas. I regret to skip the ranches west and southwest of San Antonio,[83] many of which

---

83. The acreage of the large ranches of Southwest Texas has been checked chiefly from the 1948 tax rolls and acreage that belongs to the greater ranching organizations (it is hoped without error) has been added to their total holdings found elsewhere.

The ranch statistics of those Southwest Texas counties are as follows:

Val Verde County

West, Massie 35,579 acres; West, W. W. 28,072; Martin, J. H. 28,852; Mayfield, Jno. T. & Marion 16,441; Mayfield, F. T. 13,478; Lausen, Wm. 30,333; Madison, Almond 37,771; Kelly, Joe 14,361; Kelly, P. W. & Mary 13,186; Kelly, P. W., Cecelia Eleanor 9,236; Ingram, J. F. & Merle 28,007; Ingram, J. W. & Marie, 25,438; Gillis, Roger & Willie 54,836; Jarrett, E. V. & Violet 23,304; Bluff Creek Ranch Co. 20,683; Whitehead, Emma Fawcett 9,249; Whitehead, F. H. & Winnie 19,526; Whitehead, Forrest C. 18,050; Whitehead, Geo. & Jennie 29,824; Whitehead, Louis D. & Geo. B. 10,271; Whitehead, W. R. & Dora-Willie B. and Della 18,849; Whitehead, Wardlow 19,180; Wilson, B. E. & Co. 10,494; Wilson, B. E. & Elsie 41,152; Wilson & Hodge 17,841; Rose, ABB Est. 7,167; Rose, Joe Therrell & Llewellyn 3,920; Rose, Martin Jr. 4,058; Rose,

are large enough, if placed in the populous East, to seem like a little nation pushed in among the farms and villages. Such properties as the Duval County Ranch with its 118,000 acres, the McFaddin Trust of Jefferson and Knox counties with some 140,000 acres, the holdings of McGill Brothers and numerous others deserve mention certainly as much as many who have already been noted, but there is a reasonable limit to the length of any chapter. The divided empires, such as the old 200,000 acre Jones Ranch of Jim Hogg and Brooks counties and the parts of the historic and large O'Conner Estate of Victoria and Goliad counties, not to forget the remaining acres left by old Shanghai Pearce in Wharton county—all of these belong in any complete discussion of the big Texas ranches. But this one chapter has required more than a year's search for facts, the writing of several hundred letters, and the checking of the tax rolls of nearly 100 counties, and I believe it has reached a result for the top ranking ranches that will be subject to only slight revisions. The accompanying maps of the big ranch country of Texas contain many details that cannot be told here.

But the question remains "Which are the dozen or more greatest ranches in Texas?" How would you answer that query? Would you write each rancher in Texas and ask him how many

Guida Miers 5,433; Rose, Martin & Guida 13,556; Rose, Pat & Son 12,107; Rose, Pat Sr. 10, 809; Cauthorn, Albert & Jerry 1,180; Cauthorn, John 11,119; Cauthorn, Robert & Alice 7,416; Cauthorn, Virgil & Mary Ann 14,488; Fawcett, Elmer J. 10,248; Fawcett, Mrs. F. E. 889; Fawcett, H. K. & Elke Mae 17,542; Fawcett, Lee 9,462; Fawcett, Walter 9,447; Taylor, Mary Irene & J. O. Jr. 13,442; Taylor, Mary Irene 7,436; Taylor, Lois Nell Whitehead 10,247; Davis, J. R. 12,363; Davis, L. E. & Nettie 19,234; Prosser, R. W. 17,570; Prosser & Walker 13,307; Sellers, Clyde & Winnie 12,215; Sellers, Dorothy & Clyde 5,777; Sellers, Jack M. 7,664; Miers, R. L. 32,454; Miers Bros. 33,731.
Kinney County

Whitehead & Wardlow, 17,939; Silver Lake Ranchers, Inc., 56,944; Beidler, A. F. Est. 51,306; Herbst, Mrs. LuAnna Page 6,826; Herbst, Max Page 9,333; Herbst, Fred W. Jr. 7,432; Herbst, F. W. Est. 7,080; Mitchell, Tom 20,507; Mitchell, John 9,517.
Terrell County

Prosser, R. W. 31,505; University of Texas 61,884; Pakenham, Sallie Est. 33,042; Rose, Est., 11,310; Rose, Landon & Son 13,847; Murrah, Tol 29,030; Judkins, Connie Russell 33,063; Blackstone & Slaughter 33,886; Allison, H. P. 29,736; Scott, N. M. 23,126; Russell, J. J. 35,372; West, W. E. 12,800; West, W. W. 12,800; Barksdale, Roy 17,913; Barksdale, Mrs. W. J. Est. 11,520; Mitchell, G. K. 15,551; Mitchell, J. C. 13,119; Mitchell, Mrs. Mary, 36,684.
Frio County

McClain, Marrs 11,086; Shiner, Emma F. 27,627; Oppenheimer & Lang 34,216; Halff & Oppenheimer 63,170.
Kerr County: Shreines, Gus F. 29,486; Shreines, Mrs . M. B. 53,229; Shreines, W. Scott 6,440.
Medena County: No Large Acreage.
Bandera County: McClean, Marrs 2,049.
Real County: Auld, A. D. 8,342; Auld, J. M. 14,962; Auld, William 2,101.
Edwards County

Schreiner, A. C. Jr. 10,289; Schreiner, Chas. Jr. 12,736; Schreiner, Mrs. M. B. 22,258;

cattle he has? If so, you would do well to get a dozen replies, for ranchers just do not feel the urge to answer that kind of mail. I speak from experience. The fine hospitality that still exists on dozens of ranches not very often extends to letter writing.

The next best method of ranking the big ranches, as imperfect as it is, would require us to find out how many acres a ranching organization both owns and operates. Leased acreage is not included in this final judgment, because it does not furnish a permanent basis for estimate. If two ranches even with different brands are under the same ownership, they are here counted as one ranch. Two or more different ranches owned by different members of the same family are counted separately unless they are operated as one business. Another thing, the mere fact of ownership of land does not make one a rancher; to qualify he has to "run cattle" on his land. Frazer, Burr and McAlpin own more Texas land than anybody else; but they are not ranchers, and their land does not constitute a ranch because they do not own any cattle.

Now let us get down to the business of ranking the big ranches. Let us begin with the King Ranch Corporation down in extreme south Texas. We find from the tax rolls of the seven Texas counties in which this great grassland empire is located that it actually owns 723,699.57 acres.[84] The ranch has 10,000

Silver Lake Ranch Inc. 26,620; Moody Bros. 17,759; Mayfield, Ed. C. & Son 8,264; Mayfield, Ed. C. 22,266; Wardlow, L. B. 7,680; Wardlow, L. J. 10,149; Wardlow, Mrs. Mira 7,113; Wardlow Bros. 20,055; Whitehead, F. H. 3,692; Whitehead, W. R. 5,809; Whitehead & Wardlow 42,390; Peterson, Sid C. Est. 22,048; Peterson, Joe Sid 11,076; Peterson, Hal & C. V. 22,280 (The Petersons own the Kerrville Bus Lines.)
Dimmit County

Halsell, Ewing, 28,423; West Production Co. 32,253; McKnight, S. E. Est. 27,684; Briscoe, Dolph & Georgia 49,864; Harrison, Dan J. Jr. 54,171; McClean, Marrs 30,957;
Maverick County

Halsell, Ewing, 65,435; West, J. M. 58,909; West Production Co. 2,659; Chittim, Marstella Est. 59,056; Davidson, Lynch & Co. 25,715; Saner, Mary R. 32,936; Mangum, Hal L. 33,358; Chittim, Norvel J. 29,929; Burr, J. K. Est. 77,100; Chittim, Jas. M. 34,965; Cage, Jimmie Leona 73,756; Wright, Turleta Chittim 52,295.
Zavalla County

McClean, Marrs 20,778; Pryor, D. M. 18,779; Pryor, Ike T. Jr. 33,731; Chittim, Norval J. 21,579; Norton, Richard W. 71,378.
Uvalde County: Kincaid, F. T. 25,787.
La Salle County

Burns, T. E. 23,383; Frost, J. H. 37,183; Martin, J. C. 11,425; Martin, R. V. 1,641; Martin, Albert 11,260; Martin, Antonia M. 10,191; Nueces Land & Livestock Co. 28,987; Parr, Geo. B. 34,268; Silver Lake Ranches 16,302; Texas Syndicate 75,426.
Webb County: Callaghan Land & Pastoral Co. 131,250.

84. The 1947 tax rolls show that the King Ranch Corporation had 410,209.41 acres in Kleberg County, 28,477.17 adjoining acres in Jim Wells County, and a communication from the tax collector's office shows that it had 20,460.90 adjoining acres in Nueces County. The 1948 tax rolls show that the ranch owned 198,618.93 acres in Kenedy County and the 1947 rolls showed 656 acres in Brooks County and 46,544.10

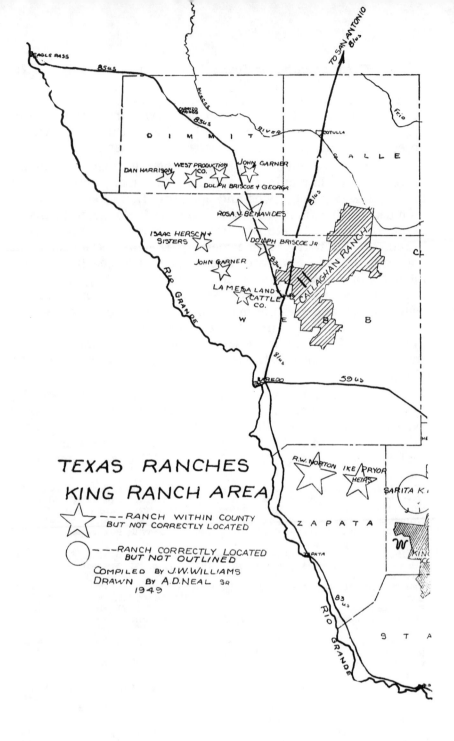

TEXAS RANCHES
KING RANCH AREA

⭐ --- RANCH WITHIN COUNTY
    BUT NOT CORRECTLY LOCATED

◯ --- RANCH CORRECTLY LOCATED
    BUT NOT OUTLINED

COMPILED BY J.W.WILLIAMS
DRAWN BY A.D.NEAL SR
1949

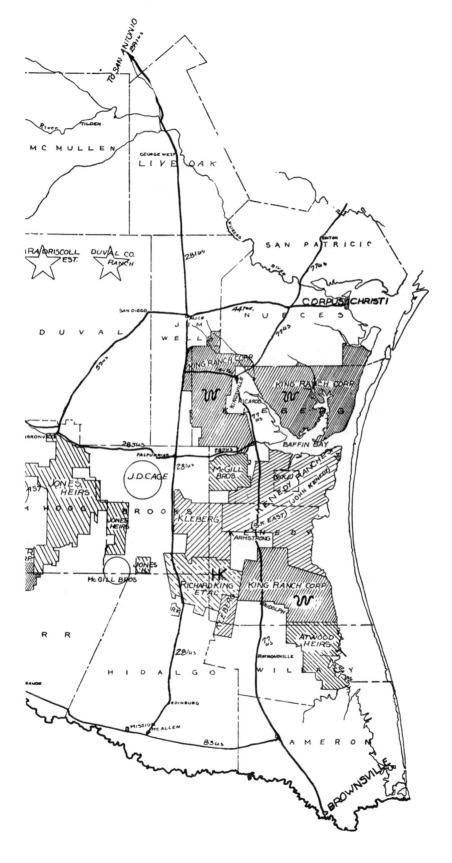

additional acres in Pennsylvania, which brings the total up to 733,699.57 acres.

Now I'm sure the reader has read repeatedly that the King Ranch is a million-acre spread of grazing lands—not just the near 735,000 acres shown above. Let me explain. The original King Ranch which Richard King had built up to cover 500,000 acres at the time of his death in 1885 was expanded to more than a million acres under the management of King's widow and Robert Kleberg senior, his son-in law—some estimates place the acreage as high as a million and a quarter.[85] But Mrs. King died in 1925 and in her will provided for a division of the property some ten years after her death. Before the ten years had expired, the elder Kleberg died, and his son Robert Kleberg, Jr. came to the head of affairs.

In due time the division of the ranch took place, and some parts of it were put back together and known as the King Ranch Corporation under the leadership of the younger Kleberg. Side-stepping any attempt to delve through the various provisions of Mrs. King's will, the 1947 and 1948 tax rolls show the break down of the once immense King Ranch as follows:[86]

King Ranch Corporation	723,659.57 acres
Robert J. Kleberg, Trustee	130,878.78 acres
Richard King of Corpus Christi and Associates	160,608.15 acres
The Attwood heirs	130,003.77 acres
Total	1,145,190.27 acres

There are some other lands in Jim Hogg County that were once a part of the King Ranch but not accounted for here. There may be included in the acreage above small amounts that were not part of the original King Ranch, but these presumably are not large enough to greatly change the picture of the great ranch

acres in Jim Hogg County. A letter from the tax collector shows 18,763.06 acres in Starr County.

85. See "The World's Biggest Ranch" **Fortune**, December 1933; also see **Time**, December 15, 1947. The above magazine articles give much of the history of the King Ranch. The present acreage has been determined, however, from the tax rolls.

86. The lands of which Robert J. Kleberg, Jr. is trustee are 52,576.27 acres (1948 tax rolls) in Kenedy County and 78,302.51 acres (1947 tax rolls with a possible error in compilation) in Brooks County. Richard King and associates have (tax assessor's letter) 78, 775.6 acres in Hidalgo County, 47,383 acres (1948 tax rolls) in Kenedy County, 29,962.65 acres (1947 tax rolls) in Brooks County, 1,167.9 acres (1947 tax rolls) in Jim Wells County, 3,320 acres (1947 tax rolls) in Starr County and possibly other acreage. W. E. Atwood et al is credited with 108,855.3 acres (1947 tax rolls) in Willacy County, 10,316.07 acres (1948 tax rolls) in Kenedy County and 10,832.4 acres (1947 tax rolls) in Jim Hogg County.

founded by Richard King a hundred years ago. It is obvious that the present King Ranch Corporation of nearly 735,000 acres is a smaller organization than the old original. Now, the question arises, did this paring down of the great King-Kleberg grassland empire take it out of first place among American ranches? The answer is yes and no both according to the date one has in mind.

Let us compare this present-day King Ranch with the Matador Land and Cattle Company as it existed up to 1951. The tax records of Motley, Cottle, Dickens, Crosby, Floyd, Oldham and Hartley counties showed a total of 790,321.28 acres owned by this big ranch. Also the company had a 17,572 acre ranch in Montana.[87] The total was 807,893.28 acres. From these figures it is evident that the Matador Ranch owned more than 70,000 acres in excess of that of the King Ranch Corporation.

It should be added that the King Ranch Corporation operates, by grass lease, the 130,000 acres of which Robert Kleberg is trustee, and enough additional leased land to run their total up to 900,000 acres—but it should be added also that the Scotch owners of the Matador Ranch long held a lease on the half-million acre Belknap Indian Reservation in Montana and enough additional grazing rights in that state to bring their leased lands alone up to around a million acres. If leased land is to be included, you can mark up the score for the Scotchmen at 1,800,000 acres.[88] Thus it is apparent that America's largest ranch, whether you count acres in fee simple or acres by lease, was for a time none other than the giant Matador Land and Cattle Company. This big Scotch syndicate was well in the lead for a full decade. At most— up until 1946 when they began to sell off a fringe of outlying tracts—these Scotchmen held title to some 875,000 acres in addition to their million acres of grazing rights in Montana.

Now, mind you, no attempt has been made here to determine the most valuable American ranch. If the great ranches are to be weighed according to value, the vast oil wealth of the owners of the Waggoner Estate might tip the scales in their favor. We turn away from any such appraisal because the many factors of value are too nearly intangible to weigh—but on the other hand acreage is a matter of record. When it comes to acreage—despite several magazine articles to the contrary—the long race for the

87. John Mackenzie to J. W. Williams, August 25, 1947.
88. The **Dallas Morning News**—an undated clipping published some time during 1948. James Wright of Dundee, Scotland, a director of the Matador Land and Cattle Company Ltd., disclosed in an interview with Frank X. Tolbert of the News the approximate acreage and the number of livestock held by the big **Matador Ranch**.

lead among the nation's big ranches undoubtedly changed to the Scotchmen when the great King Ranch was broken up.

And such was the picture up until 1951. Then as we have seen, the Scotchmen sold out and dropped out of the lead just as the XIT Ranch had done a half century before them. Such a move was not anticipated when the earlier chapters of this volume were written. This new turn of events places the King Ranch again at the head of the list.

Now let us attempt to tabulate the big ranches that are still intact. Leased land will not be included because of its temporary nature. It is not improbable that even the acreage owned in fee has undergone some changes since the mountain size task of gathering these figures was completed. Thus admitting the possibility of imperfections, the list of big Texas ranches follows: [89]

(1) The King Ranch Corp. _____ 733,699.57 acres

(2) The W. T. Waggoner Estate _____ 510,000.   acres[90]

(3) The West Cattle Co. _____ 502,199.90 acres

(4) The Burnett Ranches _____ 449,415.75 acres[91]

(5) The Kenedy Ranches _____ 436,521.65 acres[92]

(6) The Gage Ranches _____ 398,313.62 acres

(7) The Cornelia Adair Estate _____ 319,139.50 acres

(8) The Swenson Land and Cattle Co. ____ 297,848.   acres

(9) The Reynolds Cattle Co. _____ 275,531.07 acres

(10) W. L. Ellwood Estate _____ 268,649.87 acres

(11) Lykes Brothers _____ 264,554.95 acres

89. Previous notes have disclosed the source of most of the acreage figures given in this table. The source of the remaining ranch areas will be added in notes that follow. It must be admitted that some of the large ranches may hold acreage out of Texas that has been overlooked. Such additional information may change the order of magnitude of these great ranches. Efforts have been made to hold such errors to a minimum.

90. The acreage of the Waggoner ranch—510,000 acres south of Vernon, Texas, and 176,000 acres purchased from the old Bell Ranch northeastward of Las Vegas, New Mexico but now operated as the Guy Waggoner Estates—was furnished by R. B. Anderson, former manager of the Waggoner properties.

91. The acreage of the Burnett ranches comes from two sources. The judgment to Mary Couts Burnett (Vol. 424, pp. 33-38, Wichita County Deed Records) reveals the acreage in the S. B. Burnett ranches. The inventory of the estate of T. L. Burnett Vol. 30, p. 203, Wichita County Probate minutes) shows the area of the T. L. Burnett ranches. The acreage is as follows: the King County ranch 208,349 acres, the Carson County ranch 107,502 acres, the Wichita County ranch 26,499.3 acres and the combined ranches in Cottle, Foard and Hardeman counties 107,065.45 acres.

92. The 1947 or 1948 tax records of Kenedy and Jim Hogg counties show that Sarita K. East owns 194,256.50 acres in Kenedy County, and 43,784.76 acres in Jim Hogg County and that the late John Kenedy owned 191,570.39 acres in Kenedy County—also that John Kenedy, Jr. owns 6,910 acres in Kenedy County.

(12) Callaghan Land and Pastoral Co. .... 218,500.   acres[93]
(13) H. L. Kokernot and Son .... 217,155.90 acres
(14) Sawyer Cattle Co. .... 172,492.40 acres
        (Also a ranch near Santa Fe, New Mexico)
(15) Pitchfork Land and Cattle Co. .... 195,000.   acres[94]
(16) Ewing Halsell 166,793   acres
        (Also a ranch near Tulsa, Okla.)
(17) Elsinore Cattle Co. .... 186,394.67 acres
(18) Fowlkes Brothers .... 167,025.   acres
(19) W. J. Lewis .... 156,000.   acres

Clarence Scharbauer is reputed to own more than 150,000 acres but part of the figures are not available.

Probably there are other holdings of 150,000 acres or more listed in different tracts that are in fact under a single ownership without any evidence of such fact in the tax rolls.

Some of these ranch statistics need a note of explanation. The W. T. Waggoner Estate for instance has 510,000 acres in its big pasture south of Vernon, Texas, and until recently 176,000 acres east of Las Vegas, New Mexico that is now held separately as the Guy Waggoner Estate. In regard to the West Cattle Company, it is assumed here without positive confirmation that this large livestock business grazes the lands of West Securities Company, J. M. West and West Production Company.[95] The Callaghan Land and Pastoral Company has more than 30,000 acres under lease in addition to the land owned in fee. In Brewster County, relatives of H. L. Kokernot and Son own 86,979.28 acres of land. If this amount be added to the larger Kokernot property listed above the total is 304,135.18 acres. The Ellwood Estate does not include the lands of Ellwood and Towle and Miss Patience Ellwood, but it does include 126,000 acres near Tucumcari, New Mexico. Obviously corrections or future changes in the grouping of lands that operate together as one unit might change the order of magnitude of this list of nineteen large ranches. Also if

93. Paul I. Wellman, **The Callaghan Yesterday and Today**, 17. This 82 page booklet packed with much well written information about the Callaghan Ranch supplemented with dozens of ranch photographs was furnished by Joe B. Finley, manager.

94. Prior to 1947 the Pitchfork Ranch covered an area given by Margaret Elliot ("History of D. B. Gardner's Pitchfork Ranch of Texas, "Panhandle Plains Historical Review, 1945, p. 27) in round numbers at 120,000 acres. The 1948 Dickens County Tax records show 44,418.53 acres as new holdings of the Pitchfork Company. In a recent interview and letter Jim Humphreys, assistant ranch manager, gave the acreage of the Texas ranch at 163,000 and 32,000 acres in another ranch owned by the same company in Wyoming. Thus the total is 195,000 acres.

95. All three of these ownerships are listed in the tax rolls of the various counties as having the same address in Houston, Texas.

some of these cattle organizations own other land which I have not discovered ,the result may change the order of magnitude.

Probably it has been explained sufficiently that if the lands of the Bivins heirs, the Shelton heirs or the Masterson heirs in the Panhandle, or the Jones heirs or the O'Conner heirs in South Texas should in each family group be considered as one property, any of these family holdings, if so combined, should be rated among the above list of nineteen large ranches. The Sugg heirs near San Angelo and the Faskin heirs near Midland also own enough land to be included in the list, but they lease most of their grass to others. Nearly as large as these great ranches are the holdings of the McFadden Trust of Beaumont, J. S. Bridwell of Wichita Falls and a number of others.

Going back to the largest of these bodies of grassland for a moment, another type of comparison might prove interesting. The Waggoner Ranch south of Vernon of 510,000 acres is the largest ranch in one body in Texas. R. B. Anderson, the former manager, says it is the largest in the United States, and I have no evidence to the contrary. Next in size is the principal body of the King Ranch near Kingsville. It covers 459,117.48 acres. Next until the recent sale was the Matador's pasture of 395,470.28 acres at the town of Matador and next was the pasture of 394,751.20 acres northwest of Amarillo that belonged to the same company.

The fifteen largest ranches in Texas that are made up of one body of land are as follows:[96]

(1) The Waggoner Ranch (south of Vernon) 510,000.   acres
(2) The King Ranch (at Kingsville) .... 459,117.48 acres
(3) The Kenedy Ranches
    (south of Kingsville) .... 385,826.89 acres
(4) The Gage Ranches
    (east of Alpine) about.......................... 370,000.   acres
(5) The C. Adair Estate
    (southeast of Amarillo) .... 319,139.50 acres
(6) Lykes Brothers Ranch (south of Alpine) 264,554.95 acres
(7) The Callaghan Ranch
    (north of Laredo) .... 218,500.   acres
(8) H. L. Kokernut and Son
    (north of Alpine) .... 217,155.90 acres

96. The sources of this tabulated information, nearly all of which came from 1947 or 1948 tax rolls, has been included in previous notes.

(9) The S. B. Burnett Estate
(west of Seymour) ......208,349. acres

(10) West Cattle Co.
(south of Fort Stockton) ......198,752.18 acres

(11) The Elsinore Cattle Co.
(southwest of Fort Stockton) ......186,394.67 acres

(12) Reynolds Cattle Co.
(west of Fort Davis) ......180,482.82 acres

(13) Sawyer Cattle Co.
(west of San Angelo) ......172,492.4 acres

(14) Fowlkes Brothers (south of Marfa) ......167,025. acres

(15) The Pitchfork Ranch
(southeast of Matador) ......163,000. acres

We have thus in this somewhat extended chapter discussed the great and the near great among Texas ranches. We have found many large ones and have discovered one great land ownership (not a ranch) large enough to cover two states like Rhode Island and both Chicago and New York City besides. We have traveled into far west Texas where even the small ranches are so large that a mile seems almost too little to count. But if we add together all these large bodies of grassland, including the holdings of the University of Texas and even the Big Bend National Park, the grand total is considerably less than one tenth of the area of the State of Texas.

Probably all of the big ranches of the United States added together would likewise make a relatively small percent of the total land area. Texas and the other two southwestern states of New Mexico and Arizona have half the ranches of the whole nation that are more than 20,000 acres in size—but let us hasten to add that this fact was deduced from the volumes of the *United States Census of Agriculture*[97] *of 1945* which unlike the discussion in this chapter, lists a ranch (actually they call it a farm) as all the lands which one management operates whether the land is leased or owned in fee. Plainly ranch statistics based on this combination of owned and leased land and ranch statistics based

97. This publication devotes (with a few exceptions) one volume to the agricultural statistics of each state. In some instances the data covering two or even three of the less populous states are included in a single volume. It has been necessary to consult all of these volumes to compile the ranch statistics included in the remaining pages of this chapter. Separate references to the various bits of information are not given here. Indeed separate references would have resulted in very extensive and monotonous listings since in many instances a single fact has been compiled from a dozen or more of these census volumes. I am indebted to the Fort Worth Public Library for the use of this United States Census of Agriculture of 1945.

solely on ownership are not the same things, but figures based on either viewpoint may prove enlightening on the distribution of large ranches in the United States.

From questioning a number of well informed cattlemen, I am convinced that no other ranches as large as the King and the Matador exist in the whole nation, but the 1945 census, regardless of the viewpoint from which it was compiled, will lead us to a much more complete understanding of the big ranch picture of the country as a whole than any such individual questioning.

From this census we can tell the story rather quickly. In the United States as a whole there are (or were in 1945) 262 ranches of more than 100,000 acres each. Nearly two-thirds of these are in the states of Texas, New Mexico, and Arizona. The nation has 503 ranches from 50,000 to 100,000 acres in area. Ten more than half of these are in the same three states. The country has 207 ranches of between 20,000 and 50,000 acres, but these same three southwestern states fall 80 short of having half of them. Texas alone has roughly one fourth of all the ranches of the various sizes from 20,000 acres up.

Next in order below these three southwestern states are Wyoming, Montana, California and Colorado. Add these to Texas, New Mexico and Arizona and the combined seven states have 87% of the ranches upward of 100,000 acres in the whole nation. They have 80% of the ranches above 20,000 acres.

In fact so completely are the big ranches concentrated in the West that all but six of those classed above 100,000 acres come from farther west than Kansas City, which incidentally is famous for a certain brand of sirloin steak—and I am surprised as much as you to know that all six of those exceptions are in Florida.

East of Kansas City—if you go far enough—millions of people in the great population centers never saw a big ranch. Why should they? The nearest of these super-ranches is almost a thousand miles from New York City and not many of the medium large class are much nearer. To be sure there is one ranch just above 20,000 acres in Illinois. Another one is in Michigan, another in South Carolina, another in Alabama, and Virginia has one ranch just above 30,000 acres in size. Also Georgia has three ranches of the smallest of the above classifications and Mississippi has three, one of which covers almost 50,000 acres; but, if you forget about Florida, that is all of the 20,000 acre and above ranches (or possibly some of them are actually farms) east of the Mississippi River.

And now as to Florida, a pretty complete spread of ranches seems to cover its long narrow surface. It has eighty-four that range in area from 20,000 acres up to the half dozen that cover more than 100,000 acres each. Florida's big ranches are almost equal in number to those of some of the western states.

However, the whole picture of the beef producing business becomes a little clearer when we roughly estimate that the 2843 ranches in the United States of above 20,000 acres each, all added together, cover only about seven percent of the country.[98] A lot of beefsteak must come from the smaller bodies of land that make up that other ninety-three percent if we are to maintain a plentiful supply for future Americans. How will the average dinnertime serving of beefsteak look for a few decades hence?

In one last chapter let us attempt to find the answer.

98. This result was obtained by multiplying the arithmetical mean acreage of each class of ranches by the number of that class and adding all classes together for a grand total. This result was then compared with the total acreage of the United States. Obviously the figure, seven percent, obtained by this means is only an approximation but it is accurate enough to roughly evaluate the part of the big ranches in producing the nation's beef supply.

16—The Nation's Beefsteak Supply

And now in this final chapter—what is the outlook as to the nation's future supply of beefsteak?

In previous chapters we have had occasion to observe that the great range on the high plains has been mostly sub-divided into the smaller plots of the tillers of the soil; and that along with the rest, the nester invasion has made great inroads into the country below the Caprock that lay to the east of these high plains. In a more recent chapter we have noted that the immense grass leases on the Indian territory side of Red River gave way to farming early in the present century—also that Burk Burnett's 17,000 acre ranch in North Wichita County went under the plow and finally was transformed into the miracle that was known as the Burkburnett Boom. All in all millions of acres have gone under the plow, and it behooves us at last to investigate the effect that this and other forces may exert on our future supply of beefsteak.

Now a member of the medical profession who experiments with a new remedy may try out his concoction on a few guinea pigs before passing it on to the multitude of people. In similar fashion a chemist may put a few things in a test tube in order to reach a conclusion that applies to the whole World. Why not try this laboratory method here? The United States is too big to cover with such an intensive study. Let us cut down the area to the size of a test tube. Why not limit the examination to the land of Bud Arnett and Harry Campbell, Dan Waggoner and Burk Burnett—the very piece of earth about which most of this book was written. This is the same big ranch country through which the writer drove an old car for many hundreds of miles in order to write a little history. If anything has happened or is now in progress that should reveal the trends in the nation's beef reserves, close study of the forces that have been at work in this limited part of the cow country ought to give some strong hints toward an understanding of what has gone on in the whole nation.

To be sure, a forerunner in miniature of such a study was included in a previous chapter that dealt with the Quaker village of Estacado, but that chapter was too limited both in area and subject matter to develop an adequate forecast of the nation's beef supply. It is now time that we proceed with greater thoroughness to an examination, limited for the moment, to that part of the Southwest about which the principal part of this book is written.

To be specific, let us thus include for special study a slice of the map of Texas 270 miles long from Wichita Falls to the New Mexico boundary line and 150 miles wide—beginning thirty miles south of Amarillo and extending to within twenty miles of the town of Big Spring. This area includes thirty-seven counties and is a perfect rectangle except that Red River cuts a long narrow slice off the northeast corner and five counties whose history began a little too early for use here are omitted from the southeast corner.

If we could lay out on a football field a giant map of this section 150 feet wide, instead of 150 miles, and 270 feet long, instead of 270 miles, and sit above it in the grandstand, we might look down upon it and watch a cross-section of the highly interesting story of the growth of the cattle business in West Texas— and incidentally gain some worthwhile pointers as to the future of beefsteak as a part of the American food supply.

Let us suppose that this large scale map stretches out in an east-west direction nearly the length of the football field and that we look down upon it from a seat high up on the north side of the field. If the big map down in front of us is made to show the topography of the country, the high plains at our right on the west end of the football field will appear very smooth and flat.

This smooth area down on our right varies from 120 feet wide on the near side of the field to about 90 feet at the opposite side. Along the east edge of this flat section our map drops down a foot or two somewhat like a bench and then gradually slopes eastward all the way to its east end. This "step-down" at the east margin of the high plains represents the outcropping of the Caprock which we have had occasion to observe several times in previous chapters.

The section east of the Caprock is not smooth like the Plains, but is furrowed by a number of rivers and creeks. Across on the opposite side of the field from us are the headstreams of the Colorado River running southeastward. A little nearer to us, beginning with some draws that cross the high plains, the Brazos flows toward the east the full length of the map. Next beginning in the broken country near the center of the map the Wichita River flows a little north of east extending to the east end of the field. Still nearer than the Wichita, the Pease River follows a parallel course all the way to Red River, which latter stream makes the irregular north boundary—or northeast boundary—of the map for about one-third of its length.

The thirty-seven counties represented by this map cover just a little more than 21,000,000 acres. According to the one-time cattleman's formula of twenty acres of range to the cow, this area originally supported grass for a million cows. As these lines are written, this immense bit of pastureland has been inhabited by the white race for around eighty years. During all of that time, much of it has been given over to the grazing industry. The scene, **far less dramatic than a football** game and yet of much greater moment to human well being, has changed constantly.

Let us sit here high up in the grandstand and look down upon this slice of the map of West Texas while several of the scenes of the past seventy years are re-enacted. Let us imagine ourselves able to push a button and flash on the giant map below us at any part of the past seventy years which we desire.

First, let us push the button that will call back the frontier days of the year 1880.[1] Scattered over much of the east half of this

1. The population statistics of thirty-seven West Texas counties for 1880, 1890, 1900, 1910, 1920, 1930, and 1940 are taken from the **Texas Almanac and State Industrial Guide, 1947-1948,** pp. 129-132. The population statistics for these same counties for 1950 are taken from the **Texas Almanac, 1954-1955,** pp. 100-132. The areas of these counties are published in this latter volume, pp. 103-106. In this and several succeeding pages, population statistics or the relation of population to area are calculated from these convenient source books. The thirty-seven counties, named according to geographical position, are as follows: Parmer, Castro, Swisher, Briscoe, Hall, Childress,

immense map appear the small figures of cowboys on horseback—not dressed in glamour clothes as the movies might show them, but nevertheless with high-topped boots and spurs and broad-brimmed hats and some on occasion with red bandannas around their necks. These figures are very small, but we shall imagine them large enough to show plainly all over the map and that their miniature ponies are able to kick up real West Texas dust. Most of these cowboys mixed with a few farmers and storekeepers are in plain view on the east part of the map. Here if spaced uniformly they would stand—or perhaps at times sit lazily on horseback—only a little more than a mile apart or a foot apart on the map. The three easternmost counties of Wichita, Archer and Baylor with the small total of 1,744 persons have, in this year 1880, a greater population than all of the rest of this twenty-million-acre slice of Texas.

Between these three counties and the benchlike step that represents the edge of the high plains is a zone something like 100 miles wide—100 feet on the map—with the population unbelievably sparse. In the east half of that hundred mile zone the cowboys are sufficiently numerous to space themselves three miles apart, but in the west half they must scamper back to a full five miles from cowboy to cowboy. In Motley County with fewer people than anywhere else within this zone there are almost seven leagues of land to the man. Perhaps, as we look down upon this scene of seven decades ago, it is not improper to wonder if actually the cowboys did not wear seven-league boots in order to cover the range!

But west of this zone up on the high plains, except for the little Quaker village of Estacado and a few people living along the canyons, the land is completely devoid of both men and cows—but let the cows wait until a little later in the story.

Now suppose we push the button again and cause the year 1890 to appear on the big map beneath us. People have begun to settle this wide open country. A railroad has been built across the northeast corner of the map, and another such boon to immigration has barely missed the south side of it. Thirty thousand people have come to live in the counties that form the border of the map on the east, northeast and southeast. Here, if evenly spaced, there is one person each 700 yards in all directions. But

Hardeman, Bailey, Lamb, Hale, Floyd, Motley, Cottle, Foard, Wilbarger, Wichita, Cochran, Hockley, Lubbock, Crosby, Dickens, King, Knox, Baylor, Archer, Yoakum, Terry, Lynn, Garza, Kent, Stonewall, Haskell, Gaines, Dawson, Borden, Scurry, and Fisher.

in the hollow center of the map the cowboys still stand only a little less than two miles apart.

From the sparse population in 1880 it was evident that ranching was almost the exclusive industry in this slice of Texas at that time wherever it was inhabited at all. Now in 1890 the border counties around the east margin of this part of the state have too many people to subsist except by a much more intensive use of the soil. The high Plains have become a ranching country, except that three counties to the west of present-day Lubbock have as yet no inhabitants.

Suppose we continue to push the button and flash successive decades on the big map below us—1900, then 1910, then 1920, 1930 followed by 1940, and finally 1950. Great changes come at each push of the button. Fifty-five thousand people were scattered over the map in 1900, then 195,000 in 1910, and 275,000 a decade later. There were 445,000 in 1930, and there was a slight decrease to 440,000 in 1940, but a sharp rise to 540,000 in 1950.

In 1880 there were 7,329 acres to each inhabitant, but in 1950 there was one person to each 39 acres of land. If we stand a coca cola bottle down on the big map below us to represent each of these inhabitants for the year 1880, the bottles would stand three and one half feet apart; but for 1950 they would stand with only a little more than half an inch between them. Plainly the big slice of Texas represented by the map below us had come all the way from the pastoral stage of living down to the era of the plow.

Is there danger that beef cattle shall become extinct in this more intensive use of the land? It would throw some light on the answer to that question if we could watch in miniature the host of farmers as from 1880 to 1950 they broke up the sod all over those thirty-seven little counties on the map spread over the football field below us. However, let us by-pass this multitude of details which would soon run into a thousand pages of print and strike directly at the combined acreage of sod turned by all the farmers within this span of sixty years.

And now comes the first big surprise in the study of this invasion by the plow! Not all of the 21,000,000 acres represented in this study have surrendered to the plow[2]—not even half of them! Up at the west end of the big map the high Plains counties represent ten million acres of West Texas dirt—only half of this area is in cultivation. Most of the other five million acres are still

---

2. The cropland acreage by counties, taken from the United States Census of 1950, is published in the **Texas Almanac, 1954-1955**, pp. 214-216.

covered with grass! Below and east of the Caprock are the other eleven million acres under observation here. Not quite three millions of these acres have been turned into cropland while, in spite of all the miles of railroads and paved highways and other signs of change, most of the remaining eight million acres of grasslands still remain as grasslands. Across the mapped area as a whole, some sixty-two percent of all these millions of acres of free grass that beckoned to the cattlemen in 1880 still grow as acres of grass—only now they are not free. But, regardless of price, they are still food for cattle and are of material moment on the nation's beef supply.

However, the fact that only sixty-two percent of this immense cattle range is still in grass may rightly cause one to wonder what other factors are at hand to offset this loss of pastureland when it is remembered that there are now many millions of additional people to consume the beefsteak grown on this reduced acreage of grass. From 1880 to 1950 the population of the United States grew slightly more than 200 per cent. On the face of such facts, do we not need an increase of 200 per cent in the area of our cattle range?

Fortunately the answer is not that simple. But before entering the more serious discussion of the several cross currents of truth that contribute toward a more accurate result, let us first observe that 1880 is not a proper year on which to base conclusions. Why? Simply because in 1880 the nation did not yet require a full use of its grasslands. As an example, the 21,000,000 acres represented by the big map below us had ranging upon it only a little more than a quarter of a million[3] cows in 1880, whereas in theory it could have accommodated a whole million of them. Sometime after 1880 the United States reached the point where its grasslands were fully stocked—but even that statement is not quite an exact truth. The nation as a whole never did fully stock its range until water was made available within reach of every acre of grass a cow might choose to graze—and that condition is not universally true even now!

Hence it behooves us to hurdle over this first important milestone in the cattle history of the country represented by the big map below and base this study of trends upon some later year —some year after this whole area is more fully utilized.

3. From the **Tenth Census of the United States**, Vol. 3. In 1880, there were 2,477 milk cows and 263,782 other cattle. Only 18 of the 37 counties reported any cattle at that time.

In 1910 these thirty-seven counties were able to enumerate a total of 818,000 head of cattle.[4] 760,000 of these were other than milk cows, and for the sake of comparison, may be classed as beef cattle. By 1925 these totals had shrunk to only 538,000 and 415,000 respectively.[5] The great drop in numbers would seem to indicate that the reduced acreage in grass was beginning to take effect. But whether or not that is the real reason for the decrease, some other perhaps mysterious force from that day forward seems to have made itself felt. By 1945 the total number of cattle in the thirty-seven counties show not the expected decrease but a very marked rise in numbers. There were 991,000[6] of them, or at last, after more than a third of the grass was turned into farm land, the million cows that these counties were theoretically supposed to be able to support in the first place, became a reality. Thus after the country had begun to devote its energies and its acres to a varied industry, producing thousands of bales of cotton and vast granaries of wheat and many thousands of other livestock, its beefsteak production reached its real climax. With fewer acres the cattlemen were able to point to the unbelievable total of roughly a million head of cattle, only one-eighth of which were milk cows. The question is, how did he do it?

The answer—well, the traditional cattlemen did not do it.

A new type of beef producer had begun to elbow his way into the limelight. This newcomer is none other than the stock-farmer. The nester who had turned the sod on many of the large ranches and planted wheat, oats, and row crops had at last added a few beef cattle to his little farm. He has found a way to produce a cow on fewer acres of ground, and hence when the market demands, possibly he can rise to meet the increased requirements for beef. Even though the number of cattle on the farms and ranches of this West Texas area has slumped since 1945 that fact does not in any way erase this demonstration of productive capacity.

Let us examine this new entry into the land of beefsteak. Here let us call names and point out actual examples. W. H. Key operates a farm[7] of some 400 acres at the head of Lake Wichita southwest of Wichita Falls. He has 250 acres of natural pasture

4. From the **Thirteenth Census of the United States.**

5. **The Texas Almanac and State Industrial Guide for 1927,** pp. 130-132.

6. Report of the **United States Census of Agriculture, 1945. The Texas Volume** lists cattle statistics by counties.

7. This illustration of the productive capacity of the stock farm of **W. H. Key** and another such illustration of a stock farm operated by **A. R. (Dick) Etter** came from inquiries made five or six years ago.

grasses such as the traditional cattleman grazes and 150 acres of cultivated land planted largely in wheat and oats. Throughout much of the winter and spring, Key permits his cattle to forage on these growing crops of wheat and oats, removing his herds at the proper time in the spring to allow the grain to grow to maturity. After the grain crop has been removed—usually with a combine—the cattle are turned back on the wheat stubble to make their own living during much of the summer. During the fall and early winter the cattle are put back on grass, and as the winter increases in intensity, supplementary feeding is required. This feeding period usually lasts about seventy-five days and the average cow consumes 300 pounds of cottonseed meal and hulls, three bales of alfalfa, and four bales of lower grade hay such as Johnson grass.

This feed for each individual cow may be grown as a rule on one acre of ground, which makes it possible to check up on the efficiency of Keys' method of stock farming. Here on 400 acres of land are grown seventy-five head of beef cattle or an average of 5-1/3 acres per cow. Add to this the additional one acre per head required to produce the meal, hulls, and hay purchased for and consumed by one of these animals and we find that W. H. Key is producing one cow to each 6-1/3 acres of ground. It is important to the national economy that Key is producing and marketing a wheat crop in addition to his cattle, but it is beside the point here. Also beside the point is the fact that the meal and hulls consumed by each of his cows was only a by-product of a cotton crop where an important staple was produced for the world's clothing requirements. Here we are trying to arrive at the future outlook of beef production; and however nice it may be to know that some of the acres that produce beefsteak are doing double duty, the other commodities from these stockfarms that find a place in the American market are not at the heart of our discussion.

These same elements of dual production are evident on the stock farm operated by A. R. (Dick) Etter a few miles northwest of Wichita Falls. Dick has 900 acres of land, half in grass and half in cultivation. The cultivated land he mostly plants in wheat and a somewhat smaller acreage he sows in oats. He pastures this wheat and oat land during much of its growing season and pastures it again as stubble and waste straw after the grain has been harvested. In all, these crops furnish him pasture for his cattle at least six months out of the year. The grass land serves as an

ever ready standby and as feed for his cattle during the fall of
the year. Normally he takes care of the winter feeding problem
with baled oats. He cuts his oats with mowing machinery and
includes both the grain and the straw in the resultant hay crop
which, so contituted, has plenty of both concentrates and rough-
age for his full requirements. Etter takes care of 150 head of cattle
on his 900 acres—or one cow to every six acres of land. Except
once, because of the lack of machinery to bale his oats, his farm
has been self sufficient; and he has since procured the most highly
improved machinery obtainable for his purpose to insure against
a recurrence of this shortage of winter feed.

Evidently W. H. Keys and Dick Etter and their type of stock-
farmers can put land to a much better use than did the early day
open range ranchers. The 21,000,000 acres of land represented
by the big map down below us was in theory grass for a million
cows under the traditional method of ranching. At six acres per
animal it is food for more than three million cows under the Etter-
Keys methods of stock farming.

Undoubtedly there are hundreds of stock farms similar to
the two mentioned, for there is little else to explain how the
thirty-seven counties on the big map can at last support a million
cows with eight million acres of grass plowed under. At no time
back in the days when grass was the sole reliance of the cattleman, does this area seem to have even approachd this total of a
million cows. With the rise of this new agriculturalist who is part
farmer and part rancher, it has become a possibility. Since 1945
drouth and other factors have caused this figure to sag far below
the million mark; but whenever the demand for beef is strong
enough over a long stretch of years, there is little reason to believe that the totals may not even exceed this figure. Without
any great revolution in the crop system now in vogue, it is not at
all impossible that the total may some day reach the two million
mark and even go beyond.

Wheat and oats are an important part of much of the present-
day stock farming in this area. Dick Etter raises a cow on a
pasturage of three acres of this kind of grain supplemented by
three acres of grass. One and a half of the twenty-one million
acres considered here are now planted in these grain crops. Put
with this grain acreage one and a half million acres of grass and
you have according to Etter's plan of farming a living for a half
million cows and a wheat crop added to the bargain. Certainly
no set ratio of the number of stock farm acres to the cow can be

established but the fact of an increased beef production per acre is hardly subject to question. If wheat or other pasturage and some feeding can be added to grass, a new answer to the problem is to be expected.

Wheat farming is definitely on its way up. The thirty-seven counties under observation here have about as much wheat acreage now as all of Texas had in 1920. It was in the early 1920's that cattle production in this area reached its lowest point within this century. Many of the big ranches had been broken up and no new agency seemed at hand to take up cattle production. It was then that stock farming as a considerable force had its zero hour. It was a force ready to be felt. As the wheat acreage grew, so did this new method of producing a cow on fewer acres. Wheat production in Texas as a whole has increased some fourfold since 1920, with a parallel rise in our particular area. Apparently it has had something to do with this rise in stock-farming.

There is much additional land suitable for wheat farming, and along with it much room for expansion of stock-farming. However, one should not expect to limit stock-farmers to the wheat-oats-grass formula. The big spot on the map of West Texas under discussion here, during one year, produced some 400,000 tons of cotton seed; and it has approached that figure during other years. When properly processed by oil mills, this alone is food for a great part of the cattle of this area—although many car-loads of it are shipped to the cattle country that lies outside the cotton belt. Other processed foods are widely used in cattle feeding.

Obviously many gallons of water have poured under the bridge since 1880 when the twenty-five pioneer ranchers met on Chimney Creek. Then free grass was the one and only feeding formula known to this new cow country. Now one looks in vain to find a ranch that does not do some supplementary feeding in winter—it saves the lives of many cattle that must otherwise be charged off the books as a total loss. Both ranching and the newer stock-farming have gone a long way during the seventy-four years since the little meeting on Chimney Creek.

Along with these changes has gradually unfolded one of the strange tales that is related to the cow business. While the cowboys of the rip-roaring 1870's and 1880's drove Texas cattle over the long trails to Kansas, far-away Africa[8] was making a seemingly insignificant contribution to the land of cattle. In 1874 some seeds of Jerusalem corn or "gyp" corn were brought from Egypt to California. Some fifteen years later this strange new plant reached Kansas. Two years after the arrival of these seeds in California, America held its first centennial celebration in Philadelphia. Some odd-looking samples of sorghum seeds, that might be classified as red and white kafir corn from Orange Free State, were put on exhibit at this Philadelphia exposition. General Gray of the British Army, who visited the "Centennial", brought some of these novel African sorghum seed to the Commissioner of Agriculture of the State of Georgia. The Commissioner parceled out his new find in very small packages to various planters within his state. About a thimbleful of the sorghum seeds reached H. F. Watson of Campbell County in February, 1877. Watson planted them and continued to re-plant each year until 1885 when he turned over some of his greatly increased product to a seed firm in Augusta, Georgia. The firm advertised widely and began to sell packages of seed as far away as Texas and Kansas.

Some of the new type sorghum was planted in Kansas in the year 1886 with enough success that a seed firm in Lawrence began to distribute the new blessing in its part of the semiarid west.

In 1887 a farmer in Ellis County, Texas, planted five kinds of seeds which he had acquired from the firm in Augusta, Georgia —kafir, milo, African millet and two other varieties. He planted them alongside each other and, according to some accounts, he unwittingly produced the first of the blackhull kafir hybrids. To say the least, he had an interesting new product that in some respects foretold the day of greater grain sorghum production.

In addition to the "gyp" corn that came from Africa by way of California and the red and white kafirs that came out of Watson's thimbleful, the Garden City, Kansas, Experiment Station received independently some seeds which the United States Department of Agriculture had received direct from Africa. Perhaps there were other sources of the grain sorghums.

But, regardless of the number of sources, this new kind of row-crop began to spread over thousands of plains country acres

8. The story of the sorghums, to be found on the next few pages of this book, comes from a series of articles by T. C. Richardson in the Farmer-Stockman. Mr. Richardson's first articles "The Saga of the Sorghums—An Immigrant Makes Good" was published in August 1945. The other articles on the subject were published in the same journal in September, October, and November.

during the 1880's and 1890's. In the dry year 1891, a fifty acre field of Jerusalem corn produced forty bushels per acre. It looked like the answer to the horse and mule feed problem in this new country that was too dry for corn. It even had possibilities as cow feed.

What happened between the days when the sorghum seed from Africa were introduced in the West and the early 1900's is largely a forgotton story. Several unknown farmers, either by design or by accident, produced some pretty good hybrids that needed only to be distributed—then came the agricultural experiment stations and the men of science. It was this new group of unsung Americans, the agricultural agents, who put the final touches on one of the most valuable gifts ever passed from one continent to another. Out of Watson's thimbleful and the other miscellaneous seed packages these tireless workers did much toward building a firm foundation under the ranching and stock-farming economy of the plains.

Only a little of these revolutionary accomplishments can be told here. It was 1903 when the U. S. Department of Agriculture sent A. H. Leidigh to Amarillo to open an agricultural experiment station. At first Amarillo failed to furnish the land but Al Boyce, manager of the mammoth XIT Ranch, told Leidigh to come to Channing some 40 miles away and help himself to land for his experiments. Leidigh had received a wide variety of seeds from his department and even rode around on muleback to find other varieties for himself. He found a dwarf type of milo called Dawn that stood the drouth at Channing better than anything else, and proceeded to propagate it and broadcast the seed to the farmers. It became a very popular forage and grain crop—one that helped greatly in the nester invasion of the plains.

In 1905 A. B. Conner at Chillicothe, Texas—another of the experiment station men—bought a crop of dwarf milo from Judge J. F. Bradley of Memphis, Texas. Nobody knows when or where this "Memphis milo" on the Bradley farm originated, but it hardly mattered to Conner. He and some of his colleagues in other stations were quick to increase this valuable seed supply until many thousands of West Texas acres were soon to produce it by the thousands of bushels.

Along with this early production of milo came another strange type of sorghum from Africa. It was 1909 when A. B. Conner of the Chillicothe station received the first handful of Sudan

grass seed know in North America.[9] It was not long until this
small quantity had multiplied enough at his station for distribution to the public and soon thereafter it had become a popular
crop. The advantages of Sudan in a stock-farming area are evident. It can be planted in rows and cultivated until it is large
enough for forage—then old bossy and her associates can be
turned in to graze and grow fat. Under favorable conditions
wheat may be pastured about four months of the year, chiefly in
the dead of winter. This leaves a stock farmer with the problem
of finding grass or feed for his cattle during the remaining eight
months of the year. His crop of Sudan grass can help to fill that
void. It has proved to be better adapted to the grazing requirements of the stock farmer than nearly any of the other sorghums
—but it has certain faults.

It is subject to charcoal rot and bacterial leaf spots and does
not mature its seed well in the late fall. Equally bad, if the sudan
crop is threshed for its seed, the straw has little value as cow feed.

In 1934 R. E. Karper of Lubbock and J. Roy Quinby of Chillicothe set out to see what could be done about these defects. By
1937 they were able to announce a cross between Sudan and
Leoti, one of the sweet sorghums. They called the new forage
plant Sweet Sudan. Patiently they planted and worked and
waited for another half dozen years before they were willing to
offer their new hybrid to the public. It was 1943 when they
divided nine pounds of the seed into tiny four ounce packages and
distributed them for actual production. The new Sudan became
a sensation from the beginning. When planted in rows alongside
other rows of ordinary Sudan, cattle grazed the sweet variety
down to the ground before they seemed at all interested in the
older type. After three years Karper estimated that some
15,000,000 to 20,000,000 pounds of the seed would be available for
the following season. One planter had increased his four ounce
package to 100,000 pounds by the end of the second year.

Sweet Sudan, it has been discovered, is not greatly affected
by the diseases common to the older type of forage plant. It stayed
green in the fields for perhaps 30 days longer than the original
grass from Africa and even the straw is good feed. There is little
wonder that the new Karper-Quinby hybrid has spread from New
Mexico to Harper's Ferry and from Montana to the Gulf of Mexico.
It replaced half of the sudan acreage of the United States during

9. From a letter from Senator George Moffett to J. W. Williams dated February
11, 1948.

the first three years after Karper and Quinby had passed out some three dozen four-ounce packages of the seed.

Grazing tests have shown excellent results for the new product. The Blackland station at Temple, Texas, found that steers turned in to pasture on Sweet Sudan had gained a total of 367 pounds per acre at the end of an 83 day test. That is a full fifteen times as much as the weight gain expected from a whole year on an acre of natural West Texas grass according to the old 20-acres-per-cow formula!

Another sorghum that began its North American career at the Chillicothe station is Hegari. Like Sudan grass it had its beginning in 1909 under Conner. It has had great seasons— great as a grain product and great as a forage plant—but has been erratic because of a peculiar characteristic. It is known as a "short-day" plant—actually it reaches maturity faster in short days than during long ones. Early Hegari and Bonita are two improvements on the original plant that shortened the growing period of the parent sorghum and did much to stabilize the product. Bonita was developed under Quinby at Chillicothe and introduced in 1941. It is a plant of great promise.

A word about the several types of Kafir should bring these details of the grain sorghums near a conclusion. R. E. Karper of the Lubbock station introduced Blackhull 153, which was a highly worthy product, but he "hit the market on the nose" in 1925 when he introduced Texas Blackhull. This new development soon became the leading kafir of them all and occupies an important spot in grain sorghum history. Strangely enough Karper produced it from a sorghum head sent him in 1916 from Falfurias near the southern tip of Texas.

Thus did the various grain sorghums in numerous ways and under the kindly tutelage of several able agriculturists make for themselves a large place in the cow country of southwest United States. Then almost without warning several difficulties arose. Milo was hard hit by the pythian root rot. At one time it appeared that this plant had just about as well fold up its tent and go back to Africa. It was withering in the fields and the disease was in a fair way to spread across the entire sorghum belt. Then Karper and Quinby went down to Runnels and other nearby counties into the heart of the root rot infested country. They went out into the fields where the damage was greatest—and made a discovery. In some of the very fields where the milo crop was under the heaviest punishment, there were a few healthy heads of grain. The two scientists gathered 800 good heads of milo from this center of the root-rot area and went back to Chillicothe and Lubbock. Back at their stations they planted and replanted the good seed and in record time developed a strain of milo that could stand up and grow vigorously in spite of root-rot. They distributed the new seed all over the milo country and whipped the destructive disease just when it seemed sure to make a clean sweep of this highly important dry land substitute for corn.

One of the problems of the grain sorghum country had been met squarely and effectively. Resistance to disease was perhaps the most important of these but there were other requirements of a good grain sorghum plant; varieties with short stems that ripened well and stood upright adaptable to the use of combines were needed; varieties that matured early for certain climatic areas; varieties that would make good silage; and along with all other requirements all of the varieties must be resistant to drouth. The scientists of these agriculture stations as well as those in the agricultural colleges set themselves at these tasks and produced a pretty good set of answers. So well in fact have they met the needs of the farmers that the whole grain sorghum production has been turned up-side-down within the past two decades. The leading varieties found on farms today were unknown outside the agricultural experiment stations fifteen to twenty years ago!

To the city dweller who knows quite well the function of a T-bone steak, the names of these new plants in the sorghum country are just words. However, even he is interested in the little handful of agriculturalists who have brought about this revolution. The few whose names have been mentioned in this account are some of the front rank of that little group of men. There are a number of others who rank high in these grain-sorghum researches—omitted here partly because their stories lead us too far afield and partly because their own sportsmanlike modesty makes it rather difficult to give credit to whom credit is due. To ask one of these men to relate his share in a certain discovery is like asking the captain of a hard fighting football team to point out the man actually responsible for a certain touchdown.

The job of giving full credit to all this group of scientists in experiment stations and agricultural colleges is beyond us here—but let me make a suggestion. The next time you pass an agricultural experiment station accord the workers there a silent salute for the revolution they and their kind have wrought in the semi-arid West. And the next time you pay taxes and "want to cuss the government" remember that a little fraction of your tax money goes to men like Karper and Quinby, and it will be a lot easier to smile.

But let us return to the grain sorghums that men have improved so wonderfully and especially to the relation of those new sources of food to the supply of edible beef. Go back for a moment to the little old forgotten Quaker town of Estacado northeast of Lubbock, where Paris Cox performed the first of the great West Texas argricultural experiments when he demonstrated that farming on the High Plains was a possibility. Now imagine yourself in an airplaine ten thousand feet high over that one time important village. If the time of your visit happens to be mid-summer, from your perch two miles in the air you can see, or at least you are in a position to see, one-third of the grain sorghum crop of the United States. In a recent year the thirty-seven counties which we are considering for the moment—roughly the area beneath your airplane—produced 63,000,000 bushels from the grain sorghums.[10] That immense pile of grain mixed with enough cottonseed-meal to make it a balanced ration would produce 500,000,000 pounds of beef or a little less than one twentieth of the nation's present annual requirements.

As today's market for feed runs, not much of that big pile of grain is fed to cattle. Most of it is ground and combined with other elements into chicken feed. But grain sorghums, which the experiment stations have shown can be successfully produced from Canada to the Gulf, are the nation's great standby for the fattening of beef cattle if the now ample corn crop fails to expand to meet future needs. The experiment stations at Spur, Big Spring, and other places have demonstrated that grain sorghums are pound for pound equivalent to corn in finishing beef cattle for the market. Just file that fact away for use when the market demands it.

Meanwhile the sorghums as a whole are a vital part of present ranch and farm economy. Only a part of the rather extensive sorghum crop is threshed for grain. The whole stalk of sorghum from the ground to the top of its head is good cow feed.

A few years ago R. B. Anderson, then manager of the Waggoner Ranch, showed me some pellets made from grain sorghums that were something new under the sun. Sorghums grown here are cut by silage machines and taken to an on-the-ranch factory which dehydrates this mixture of cow feed, grinds stalk and grain, and transforms it into the kind of pellets of which Ander-

10. The **Texas Almanac and State Industrial Guide**, 1947-1948, pp. 224-227.

son exhibited a sample. These pellets are put into grain sacks and kept until they are needed as food for cattle. The dehydrated pellets will last indefinitely if not used. The plant by which the Waggoner Ranch processes this new kind of feed is—or was at the time—the only factory of its kind yet constructed.

Will the time come when community ventures will make this kind of processing available to the little stock farmer? Whether it does or not, he need not sit down and wait. He can dig a trench silo and make his sorghums add plenty of pounds to his growing cattle—or with less efficiency he can simply bundle the feed and store it until it is required. This last suggestion is a common practice, but the silo is gaining in popularity. But whether by silo or by bundle, the sorghums are one of the greatest reasons why the 21,000,000 acres selected here for special study can now support a million cows after 8,000,000 acres of its grass have been turned under.

Perhaps we have reviewed enough of the elements that have made possible this increased beef production in the thirty-seven counties that have been cut from the map of West Texas and observed like a guinea pig or like a few chemicals in a test tube. We have noted the rise of the stock farmer, and the new resources that have been brought from Africa and improved by scientists to help him to succeed. We have also had occasion to look in on a little of the promising work of the experimenter, such as the testing of Sudan grass at Temple which indicated a product of fifteen times as much beef from grazing an acre of ground as obtained from an acre of natural pasture grass. Some of these laboratory examples doubtless will point the way to still greater efficiency for the stock farmer.

Improvement of production from the grasslands themselves (some of which have not been mentioned) such as the increase of the number of watering places made available for range cattle, the eradication of mesquite timber, the killing of the prairie dog, and in places the planting of new pasture grasses is of evident import without further discussion.

If every square rod of the 21,000,000 acre slice of the West under study here could be used toward beef production, making full use of all the feed that its 8,000,000 cultivated acres might produce, and if that feed were processed so as to result in its greatest possible growth of cow flesh, this large sample plot of ground would doubtless support not just one million but several million head of cattle. The balances that regulate our national economy did not permit this to occur. The cotton market made its bid for the use of this same 21,000,000 acres of ground and at most extracted from it some 800,000 bales of cotton in a single year. The wheat market made its bid and received 1,500,000 acres planted to wheat. The poultry and other market forces made their bids and received 3,000,000 acres planted to grain sorghums and a crop of as much as 63,000,000 threshed bushels. It was in competition with these and all other market forces that cattle production in this selected area nearly doubled during the past twenty-five years. The physical elements were on hand for a much greater increase had the price of beef been high enough in comparison with all other commodities.

Was the trend in this selected area typical of the trend in the nation as a whole? Has it been a fair cross section? The answer is "yes" to the first question if we mean by typical that the elements at work in the nation are about the same as those in this limited area. The answer is "no" to the second if we mean by a cross section that those elements are in the same proportion both locally and nationally. Stock farming has not been limited to one locality. Also the "know how" in agriculture, partly already in use and partly laid away in our national ice box for future use, is not restricted to any certain set of parallels of longitude and latitude. The pull of the market for or against cattle production, however, is different in different sections. In our particular selected area, the chief competitors of cattle for the use of the soil are cotton and the grain sorghums; but in a great part of Kansas, for instance, wheat is the chief competitor.

Keeping in mind this tug-of-war that goes on throughout the nation, it behooves us to check up on the actual statistics of national cattle production.[11] Back in 1880 there was not quite three-fourths of a beef cow for each person. Ten years later this ration was almost the same. But by 1900 the available beef supply had dropped until the average share per person was just a little more than half a cow. By 1910 that figure had dropped to a little less than one half, and although the high prices of the first World War had brought up the supply, there was still less than one half a cow per person in 1920. Then came the greatest drop in the nation's beef supply per person to be recorded during the past three-quarters of a century. By 1925 the average man's share was

11. Cattle production statistics may be had from census reports and various handbooks of statistics. Conveniently plotted curves of the number of cattle on farms in the United States, 1867-1948 were published in an article by C. L. Harlan, entitled "Trends in the Livestock Industry in **Texas**" in **The Cattleman** magazine June 1948.

just a little more than one third of a beef, and five years later was actually less than one-third. This slump, which in fact had hit bottom some two or three years before 1930, was the low point for all the years for which records are available.

Then the cut of beef available for the average man began to climb back a little. By 1934 it had risen to quite a fraction above one-third of a beef animal, but drouth struck at the supply on farms and ranches and the quantity went back down to almost an even one-third where it stayed until 1940. No powerful price movement was on hand to bring up the per capita ration of beef.

But it was not long in coming. World War II was just around the corner, and in its wake by 1945 the average man had at his disposal 41½% of a beef animal which is almost exactly mid-way between a half and a third of a cow. This was 90% as great as the ration per person available in 1920 at the end of World War I. It was 73% as great as grand-dad had in 1900, 58% as much as great-grand-dad had back in 1880, and although I do not trust the early statistics without some reservations, it was 75% as much as great-grand-dad or any of his neighbors had in 1870. Put these figures another way for the sake of clarity. One thousand people in 1870 had available beef supply of 540 head of cattle; one thousand people in 1880 had 720 head. In 1900 an equal number of persons had 565, but in 1920 the number was only 460, and in 1945 only 415.

Let us go a little farther for the sake of clarity. Suppose that each of these herds of cattle were driven into a corral, and we could ride by and see them. This last herd of 415 head would be by far the finest beef cattle of them all—most of them Herefords or Angus or Brahmas. Some of their bellies would almost seem to drag the ground. By contrast those older herds that came from granddad's or great-granddad's day are muscular and tough, with great horns and thin hips. Moreover so many of those older herds died on the range during the hard winters without supplementary feeding that considerably less than the 720 or 565 mentioned above ever reached the slaughter pen. Personally I would hate to trade my share of the 415 modern beef cattle for granddad's share in whatever number of his 565 actually reached the corral, or for great-granddad's share that were actually marketed of his 720. Considering the high death rate on the old winter range and the fact that a modern cow is endowed with so much more edible beef than her predecessor, the difference in figures old and recent are more apparent than actual.

But this does not mean that the quick slump in beef supply that came after 1920 was at all imaginary. That was the most severe drop in the average man's rations of beef to be found in trust-worthy statistics since the West has been a material factor in cattle history. The herd of 460 beef cattle available for each 1000 people in 1920 dropped to 310 in 1930! What is to prevent a recurrence of this shortage following 1945? It would be easy at this point to become an alarmist and to predict the worst shortage of beef in the nation's history. Based on the rate of decline from 1920 to 1930 the little herd of beeves available for each thousand persons within the nation should decline to 280 head by 1955. But will it? The truth is, a rather definite drop in supply has already occurred. From January 1, 1945 to January 1, 1947 the 415 beef cattle per thousand Americans had slumped to 385 and by the beginning of 1948 it had dropped still further to about 370.

Fear of declining beef prices must have been the dominant motive that brought about this reduction of cattle of the range. However, the very prospect of an acute beef shortage caused by this rush to market should call a halt. Prices cannot drop very far in the face of such a growing shortage. Undoubtedly, cattle raisers in general must have recognized this fact, for the drop in supply was stopped dead in its tracks during 1948. Beef cattle on farms and ranches increased from a low of 53,000,000 to 54,000,000 by January 1, 1949. This rise was fully as fast as the phenominal increase in population, so that the actual supply of some 370 beeves to an average thousand Americans is temporarily at least the low point in available supply.

More recent trends have greatly minimized the danger of any immediate shortage in our beef supply.[12] By 1952, the beef cattle on farms had grown back to 415 per 1000 persons, and preliminary figures for 1953 have apparently pushed the index much higher than that. But the rapid rise in population offers a constant challenge to cattlemen and stock farmers. To keep pace with the millions of new mouths to feed, these producers must find ways to add yet other millions to the number of American beef cattle.

And thus, the race between the number of cows and the number of people runs on and on. For many decades the contest has been a see-saw affair—there is little reason to believe it will be otherwise in the future. What the economists call the law of sup-

12. Recent cattle statistics are published in the **Texas Almanac, 1954-1955**, page **203.**

ply and demand, has kept the ratio between population and beef supply from running too far away from normal in either direction. That ratio should continue to stay within reasonable limits unless the physical elements that make for beef production cannot expand to meet future needs.

One of these physical elements—natural pasture grass—has, it is true, suffered a decline. Just as 8,000,000 acres of the 21,000,000 acres selected here for special study has gone under the plow, so has the plow made great inroads into the pasturelands of greater areas. The great strip of the West which includes the Great Plains—Texas and nine states north of it—has permitted 170,000,000 acres to go under the plow.[13] This is a deep cut into the grasslands of the cattle country, but lest one should be alarmed, it amounts to only one-fourth of the total area of those ten states. We have seen how cattle production greatly increased in our 21,000,000 selected acres while, not just one-fourth but, in fact, more than one-third of its grass was plowed under. For every handful of grass that the plow destroyed there was a new place to grow a stalk of sorghum or a bunch of Sudan grass which could produce much more beef than the handful of grass. Something less than one-fourth of the land of the United States[14] has gone into cultivation, a fact which does not make the national picture very different from the little spot of West Texas set aside for special study in this chapter.

Corn is another of the physical elements which produces beef. Iowa and other corn states add many millions of pounds to the cattle off the ranches and farms of the West before they enter the market for slaughter. In addition to all the other uses of corn a recent crop of 3,600,000,000 bushels was more than enough for this cattle feeding. If this immense crop ever becomes insufficient, the grain sorghums constitute a very large potential, ready to compete for part of the soil of the Great Plains states and supplement corn as a cattle feeder.

The general purpose sorghums and other pasture and forage crops are not a potential but an ever present force in the production of beef on the stock farms. This is the physical element that is more than taking the place of the acres of grass that have been

13. United States Census of Agriculture, 1945. The volume on farm property, (pp. 26-29) lists the cropland by states.
14. Ibid.

plowed under; it is through these agencies that the market can in time command a great expansion of cattle production on the stock farms.

However, it should be understood that grass is the cheapest source of beef yet found and that all these other types of feeding entail some labor and machinery and, consequently, added expense. On the other hand, it should not be overlooked that increased efficiency in plowing and planting cow pastures and also in harvesting these forage crops can already make heavy cuts in that expense. The pellets that are produced for cow food on the Waggoner Ranch are not touched by human hand from the time the grain sorghum is cut down in the field until the finished product reaches the sack. Surely it is out of keeping with past experience if farm machinery does not extend methods that are similarly efficient to the stock farm.

There are plenty of American acres to produce more beef than we have ever had. As long as the price of cattle is high enough to compete with other commodities for the use of those acres, beefsteak can maintain its present importance on the American dining table. We should expect that price to average somewhat higher than in time past when grass was almost the sole food for cattle—but efficient methods and efficient machinery on the stock farm may yet offer us some surprises.

* * * *

But we did not go out into the big ranch country just to find the answer to the problem of the nation's beefsteak supply. Any total, large or small, of that great staple of American food was not quite the chief objective. To be sure it has been our lot to watch the light of early morning glow into a brilliant red far up in the Texas Panhandle surrounded by vastly more than a hundred thousand tons of beefsteak on the hoof. It has also been our lot to watch the day die far beyond the salty Pecos surrounded by yet other tons and trainloads of juicy beef. We have watched some of the drama of the West as the early ranches covered the open range and as the open range was cut into ten thousand pieces by strands of barbed wire. We have watched great ranches only to see slow moving trainloads of nesters with covered wagons drive on out West and parcel them into farms.

But the farms, and the ranches, and the production of beefsteak all combined are not quite the biggest thing that was exposed to view all over the West. Perhaps the true spirit of that out-door land toward sunset was discovered in its purest form in

my interview with Red Mule Barkley out at the little town of
Clairmont.

For a moment let us go back to Clairmont and to Barkley's
story of the princely cattleman, Boley Brown—to the man who
has become a legend of the upper Brazos. Let us again drive out
on the range to the unique monument built to a man who had
spent his life among the cowboys and the cows.

Joe Meador would have gotten quite a lift out of this day's
drive. Here was the cow country as nearly in its pure form as
could be found. Here were millions of broad flat acres of grass
that had never seen a plow. Here the West stretched out all the
way from sunrise to sunset. Whitefaced cows were scattered about
in little herds as far as the eye could see. Perhaps the vastness of
it all had touched the hearts of the sun tanned men who still
applied its branding irons. Whether their hats and boots and
spurs were exactly true to those that came from the film marts
of Hollywood was of little moment. Here was a land that had
mixed with its daily routine a code of living. It had set up a kind
of unwritten Ten Commandments for its heroes. They were commandments that nobody had ever thought to put in words. Men
of the range sensed the clean cut nature of a man's adherance to
this invisible code by a kind of intuition.

Perhaps the reader will remember that it was after driving
on back roads through pasture gates and several miles across the
prairies that I finally found this strange marker to the memory
of Boley Brown. It was on a slight eminence that overlooked from
afar the quiet vastness of the Brazos River Valley.

Here out on the range, far from any human habitation, this
neat clean shaft of stone, fenced in a little enclosure and neatly
kept, pays everlasting tribute to a man—not for any official greatness nor for any acts of heroism but just for neighborliness and
fair dealing. The West that loves its fine horses with a passion
also places a premium on men who are not guilty of littleness of
soul—but nowhere else, so far as I know, has it expressed this
sentiment with a monument of stone. Here was just a shaft of
marble but the man to whom it was built was the soul of the Old
West.

Index

*See Note at end of Index, Page 307

*Note: In accordance with custom the lower case "n" following any page number in this index indicates that the item referred to will be found in the footnote on that page.

Page numbers are not printed within the photographic section of this book, but the photographs are referred to in the index by page numbers. A count and turning of pages in each case will locate the photograph desired.

Also in accordance with accepted practice the letter "f" or letters "ff" following a page number reference in this index calls attention to the fact that the desired item may be found on the page or pages following.

The foregoing notes are inserted because this book is expected to go into the hands of many persons not accustomed to dealing with such usages.